Vulnerability and Capacity Assessment to Climate Change of Malwa Region, Madhya Pradesh

Dr. Bhaskar Sinha
Shashidhar Kumar Jha

Indian Institute of Forest Management
Nehru Nagar, Post Box No-357
Bhopal – 462003 (M.P.)

Bishen Singh Mahendra Pal Singh
23-A, New Connaught Place
Dehra Dun - 248 001 (INDIA)
2017

Vulnerability and Capacity Assessment of Tribal and Non-Tribal Communities of Malwa-Madhya Pradesh

ISBN-978-81-211-0968-0

Published by Abhimanyu Gahlot for M/s Bishen Singh Mahendra Pal Singh, 23-A, New Connaught Place, Dehra Dun and Indian Institute of Forest Management, Nehru Nagar, Post Box No-357, Bhopal – 462003 (M.P.), India, Printed at Shiva Offset Press and composed by Doon Phototype Printers, 14, Old Connaught Place, Dehra Dun, India.

Preface

Climate change is a global phenomenon and has adversely impacted mankind in the last few decades. Extreme events due to climate change are unevenly distributed between nations, regions and communities and therefore the degree of impacts varies across the region. These impacts will hit the developing countries hardest and earliest, which will further worsen the existing developmental challenges due to lesser adaptive capacities of the rural poor communities. The Government of India and many Indian states have accorded top priority to the issue of climate change and developed action plan at the highest level on crucial sectors with necessary policy, regulatory and institutional framework. Developing an effective adaptation strategy would require specific vulnerability and capacity assessments for countries, regions or communities with the help of indicators. Additionally, such assessments would also help in identifying adaptation strategies that are feasible and practical for communities.

In this background, the book provides a framework for comparative vulnerability assessments of tribal and non-tribal communities of Malwa region, Madhya Pradesh. This region is considerably vulnerable due to higher prevalence of poverty, which necessitates an urgent need to scale up its anti-poverty measures and adaptation strategies. The study uses the indicator based approach, consisting of eighty indicators that are further grouped into ten components and forty-five indices, which cover major aspects of rural livelihood. In addition, principal component analysis (PCA) was used to identify and prioritize the important factors that contributes significantly to sustainable livelihood and reduces vulnerability. Along with quantifying the comparative vulnerability scores on different livelihood components for adaptive capacities of tribal and non-tribal communities of this region, the book also discusses the existing government policies/schemes/programmes which could address the specific causes of vulnerability.

The approach and framework used for assessing vulnerability to climate change in the study would help in designing sector specific adaptation strategies to strengthen the adaptive capacities of different communities in the Malwa region, and could be applied for different remote and backward regions of the country.

Acknowledgements

I am highly indebted to the Director, IIFM, for his guidance and constant supervision as well as for providing the necessary information regarding the project and also for his support in completing the project.

I would like to express my gratitude to the officials of Dhar and Ratlam District for their kind co-operation and encouragement which helped me in completing this project. I owe special thanks and gratitude to the field staff for their help in carrying out the fieldwork. My thanks and appreciation also go to Dr. Jigayasa Bisaria, Assistant Professor, IIFM, for data analysis and to those people who have willingly helped me out with their abilities.

This effort would not have been possible without the kind support and help of many individuals and organizations. I would like to extend my sincere thanks to all of them. I would also like to acknowledge all my colleagues for their critical ideological contributions. Lastly, I offer my thanks to the official staff of IIFM, Bhopal for their valuable support during this study.

Dr. Bhaskar Sinha

Contents

Executive Summary

1.1. Introduction

Human interference with the climate system is occurring, and climate change poses risks for human and natural systems (IPCC, 2014). The impact of climate change is particularly higher for some segments of society (such as poor and marginalized) and sectors (such as agriculture and forestry). Apart from this, it significantly affects the health, safety and resource availability of poor and marginalized people (Cruz *et al.,* 2007 and Stern 2007), as they are confronted with limited livelihood options due to remoteness and few incentives to stay in balance with the surrounding ecosystems. It has a multidimensional impact on humanity in terms of several socio-economic parameters like agriculture, human health, water scarcity, disease prevalence, etc. The probability of occurrence and magnitude of impacts are determined by weather and other climate related events (Brooks 2003; Goulden 2010). These interacting issues create complexities for assessing, vulnerability and hence, formulation of sector specific programme is still a big challenge for the concerned authorities.

Madhya Pradesh is the second largest state in the country in terms of area and is endowed with diverse forest resources, and a rich history and heritage. The state has clocked 24.99% agricultural growth and was bestowed the prestigious "*Krishi Karman Award*" for its outstanding performance in agriculture sector. The state is home to a large tribal population. About 36 % population of the state belongs to the marginalized Scheduled Castes (SC) and Scheduled Tribes (ST) (20.27%) (Census of India, 2011) and 48.6% of the population is estimated as living Below Poverty Line (BPL), with the rural poverty ratio (53.6 %) exceeding the urban poverty ratio (35.1 %) significantly (Tendulkar Committee Report, 2009). Many areas in the state are vulnerable with considerable poverty and the state need to scale up its anti-poverty measures. Climate change has the potential to undermine existing efforts to tackle the mountain of poverty removal in the state and the complex social development problems faced. Several studies have explicated the seriousness of recurring droughts in Madhya Pradesh, with a far-reaching and deep impact on local livelihoods

The current study is intended to understand the impact of climate change by evaluating the vulnerability of tribal and non-tribal communities in the Malwa region situated in a relatively remote location. The vulnerability in the Malwa region is high and efforts for adaptation and mitigation are comparatively weak. The region is known for its pockets of poverty and the problems of poverty are further aggravated by limited livelihood opportunities, a lower literacy rate and scarcity of water and food. The vulnerability of the Malwa region is assessed by an indicator based approach so that the vulnerable sectors could be identified easily. Further, it would be helpful for planners/policymakers in formulating adaptation programs for strengthening individual/collective assets. The study was undertaken with the following objectives:

- To identify the components of vulnerability and their impacts on livelihood of rural communities.
- To understand the perception of rural communities towards climate change, identify the primary causes and their coping mechanism.
- To identify the vulnerable social groups and evaluate community-based coping or adaptation mechanisms to the impacts of change.
- To formulate recommendations (for planners and policymakers) to improve individual and collective assets, with the aim of enhancing the adaptive capacity and resilience of vulnerable communities in their responses to climatic change.

1.2. Study Area

The study has been undertaken in Dhar and Ratlam districts of the Malwa region. The region is dominantly inhabited by the Bhil tribe (Meos, Barelas and Patelias). In addition, Meenas, Dawoodi Bohras (Shia Muslims from Gujarat) and Patidars (Kurmis of Punjab) reside in the region with remarkable variation in dialects and social life. The district of Dhar is located in the south western part of the state of Madhya Pradesh. It is located between latitude22^0 N - 22^{0}49' N and longitude - 75^0 06'- 75^0 42'

E. The district is located 266 KM east towards the state capital, Bhopal. The average altitude of Dhar District is 588 meters above sea level. The district extends over three physiographic divisions, namely, the Malwa in the north, the Vindhyachal range in the central region, and the Narmada Valley along the southern boundary. The second district of the study is Ratlam. It is situated in the westernmost portion of the state of Madhya Pradesh surrounded by the districts of Dhar, Jhabua, Ujjain and Mandsaur. The district extends from latitude 23^0 05'N - 23^0 52' N to longitude 74^0 31' E- 75^0 41' E. The total area of Ratlam District is 4861 Sq.km. which is 1.11% of the total area of Madhya Pradesh.

1.3. Sampling Strategy

The Malwa region was selected for the study because the region is known pockets of poverty and is inhabited by a higher concentration of scheduled castes and tribes. The poverty is further aggravated by limited livelihood opportunities, lesser literacy, subsistence agriculture, degradation of natural resources, scarcity of water (surface and ground), deprivation of special government programs and the absence of a strong people's organization / civil society/NGO movement in the region. In the Malwa region, Dhar and Ratlam districts were selected for the study.

Further, in Dhar district, Dhar and Gandhwani blocks were selected under the non-tribal and tribal categories respectively for the sampling. These blocks were selected on the basis of the population of scheduled tribes and literacy. The Dhar block was in the non-tribal category because of lower ST, population (36.44%) and higher literacy (57.18%), as compared to the other blocks of the district. On the other side, the Gandhwani block was selected in the tribal category because of its higher scheduled tribal (90.74%) population and lower literacy rate (34.83%). In the Ratlam district, the Jaora block (03.97%) was selected for sampling in the non- tribal category. On the other side, Bajna block comprises 95.22% tribal population and was selected under the tribal category. Under each block ten villages were selected randomly. Villages with more than 90% tribal population in the tribal rich category i.e. Gandhwani in Dhar and Bajna in Ratlam and villages with less than 10% tribal population in the tribal poor category i.e., Dhar block in Dhar district and Jaora block in Ratlam district were selected for sampling. Further, ten respondents were randomly selected from each village for personal interviews. Well-designed survey tools were used for evaluation and it helped in capturing both the qualitative and quantitative dimensions of data.

1.4. Analytical Framework

The Intergovernmental Panel on Climate Change's (IPCC) vulnerability framework breaks down the concept of vulnerability into dimensions of exposure (external disturbances); sensitivity (change brought about by exposure) and adaptive capacity (ability to adjust itself to minimize the damage). The greater the exposure or sensitivity, the greater is the vulnerability. Adaptive capacity is however, inversely related to vulnerability. So, the greater the adaptive capacity, the lesser is the vulnerability. The Vulnerability and Capacity Assessment (VCA) index is calculated on the basis of the IPCC's vulnerability assessment framework. Wherein, exposure and sensitivity together describe the potential impact that climate change can have on a system, and adaptive capacity is generally accepted as a desirable property or positive attribute of a system for reducing vulnerability (Engle, 2011). In the VCA framework, vulnerability is calculated by estimating dimensions of vulnerability (DoV), i.e., adaptive capacity and sensitivity and exposure through an indicator based approach. The DoV is composed of a set of components of vulnerability (CoV). The CoV is a set of indices that measure vulnerability by aggregating indicators. The crucial sectors of well being, that are directly or indirectly impacted by climate change is taken as CoV, namely, socio-economic profile; social network; education; health; infrastructure; natural resources; agriculture; food; water; and weather perception and calamities. Each CoV contains relevant indicators based on specificity of the region and is selected by the interactions with the peers, locals of the region, block and district level officials and a review of the literature. The framework uses primary data and employs maximum-minimum normalization to restrict the value range of indicators between 0 to 1 values nearer to 1 showing, higher vulnerability. The indicators values have been aggregated into the components and further converted into dimensions of vulnerability. These were adaptive capacity, sensitivity and exposure. The CoVs, were further tested statistically (T-test, Analysis of Variance (ANOVA) and Principal Component Analysis (PCA) to analyze the difference between tribal and non tribal blocks and factors with the highest variance.

1.5. Result and Discussion

Over the last few decades, Madhya Pradesh has witnessed a radical transformation in terms of economic and social development. The state has recorded an impressive growth in the agricultural sector (24.99 % in 2013-14) and infrastructural development, supplemented by the creation of robust institutions. The Gross State Domestic Product (GSDP) registered a double-digit growth of 11.10% at constant prices, up from 9.9 per cent in 2012-13. In contrast, the state's performance on basic human development indicators (0.375) is not impressive, and many a time fall far below the set standards of national average (UNDP, 2014). Key sectors such as health, access to safe drinking water and sanitation, housing and electricity are still causes of great concern if the state has to pursue a balanced and equitable socioeconomic development. Thus, the state has to put forth sector specific strategic efforts in improving the current pace of providing services. In the light of the above, a research study on VCA was conducted in the Malwa region of Madhya Pradesh to identify the dimensions of development for enhancing the capacity of rural communities and, thereby, the emerging policy thrust for development. The study uses indicator based approach. A total of 80 indicators were selected and further grouped into 45 indices and 10 components, which covers major aspects of rural livelihood. All the components were tested statistically (T-test and ANOVA) and they showed significant difference between tribal and non-tribal blocks, except a few.

1.5.1. Components of Vulnerability and their Impacts on Livelihood

1.5.1.1. Socio-economic Profile

The socio-economic profile of the region is measured by indicators like family dependency, education, poverty, migration, sanitation and access to electricity. Apart from this, the geographical location and accessibility to basic facilities also play a critical role in the socio-economic analysis. The component contains ten indices which are composed of 12 indicators. The socio-economic vulnerability is higher in the tribal regions and most of the indices were significantly different. The higher vulnerability of the tribal blocks might be attributed to higher dependency, poverty and migration, together with lower education status, sanitation and access to electricity.

1.5.1.2. Social Network

Social networks reflect the potential for social support. Obtaining loans (money, food, and non-food items) and assistance from formal and informal institutions during extreme events is an important strategy used by households to cope with environmental and socio-economic shocks. A social network consists of two indices which are composed of five indicators. The component showed significant difference between tribal and non tribal blocks. The vulnerability score of this component is slightly higher in the regions because of weak social networks.

1.5.1.3. Infrastructure

Infrastructural development is of crucial importance in creating better conditions of life for human societies. The infrastructural development of the Malwa region was sketched by collecting data on connectivity; accessibility to markets and banks, and availability of government resources such as primary schools and panchayat buildings. The result shows that all the indices are significantly different between tribal and non tribal blocks. The component however, do not show a significant difference because of almost similar programs/policies for development in all of the blocks. Further, the lesser adaptive capacity of tribes in terms of inaccessibility to districts and lack of awareness makes them more vulnerable.

1.5.1.4. Education

The state recorded impressive achievements for education in terms of strengthening its existing school system and innovating for reaching the unreached, increasing enrolment and ensuring widespread adult literacy in the state have led the effort. But then the state is still struggling to cope with its school dropout rates. The dropout rate is 15 % in primary and upper primary levels, and a majority of the dropout children belong to SC (17 %) and ST (43.5 %). The educational attainment of villagers in the Malwa region was captured with the help of indices like education index, youth education index and the final, which is impacted by these two indices, is the decision index. Again, the vulnerability score for education is higher in tribal regions due to higher school dropout rates and inaccessibility to higher education. The mid-day meal programme however, allures children of five to ten years to attend school. But after primary education, most of them leave education because of a number reasons such as engagement in economic activities, cattle grazing, less inclination towards education, sibling care, social beliefs and parental reluctance.

1.5.1.5. Health

The World Health Organization (WHO) claims that "the enjoyment of the highest attainable standard of health is one of the fundamental rights of every human being" (WHO 2006, p 1). The health status of the respondents of the Malwa region is measured by two indices, i.e. the health worker index and the health facility accessibility index, and each index is composed of two indicators. The overall health vulnerability score is lower in non tribal blocks because of better accessibility to community health centers and frequent visits of doctors as well as other health staff in the villages. Further, the indices were tested statistically and the health worker index of tribal and non tribal villages showed a significant difference. The overall component did not show any significant difference between Dhar and Ratlam districts.

1.5.1.6 Agriculture

A major challenge in the agricultural sector is to improve the poor quality of agriculture practised in many areas. Agriculture of the Malwa region especially the tribal blocks is challenging; it holds the key to livelihood security and rapid reduction in rural poverty. It is expected that the ongoing population growth, urbanization, and changes in climate will further reduce the availability of water resources which will affect agriculture. In Madhya Pradesh, climate change is likely to intensify and increase the occurrence of drought. As with food security, the households that are already facing water shortages are more sensitive to changes in temperature and precipitation. The agriculture of the Malwa region is further constrained by an undulating topography and labour intensiveness. The present status of agriculture of the Malwa block is captured with the help of seven indices and 13 indicators, i.e., agricultural equipment, land, cropping intensity, yield, irrigation and cropping strategies. All of these indices showed significantly higher vulnerability in the tribal regions.

1.5.1.7. Food Security

Agriculture is the prime source of food in the Malwa region although a few of them are also dependent on the Public Distribution System (PDS). To assess the food security of households, two indices and three indicators were used, a central one being the primary source of food, food sufficiency to feed all household members throughout the year, and the variation in consumption of nutritious food in the last ten years. The food insecurity was higher among tribal blocks and hence are more vulnerable as compared to non-tribal blocks.

1.5.1.8. Water Security

Water security is of vital importance to the livelihoods of the people in the Malwa block. The sensitivity of households of the Malwa region is further increased by inaccessibility of potable water. Therefore, accessibility, together with the availability, of portable water is of vital importance and should be addressed during any vulnerability assessment. Three indices and eight indicators on potable water including source; availability and accessibility of water; and time spent and distance travelled for water collection were used to measure water vulnerability of the Malwa region. The overall vulnerability score of the tribal blocks was very high as compared to the non tribal blocks. This might be attributed to the lesser number of water sources inside the village boundary and higher dependency on them.

1.5.1.9. Weather Perception and Calamities

The impacts of climate change are expected to be higher for some segments of society because of geographic location; the degree of association with climate-sensitive environment; and the unique economic, political, and cultural characteristics of particular areas. The weather prediction and calamities index was designed to capture the perception or views of rural communities on climatic variability and change. The level of education, awareness, access to weather forecasts, reliability of forecast, involvement of the formal/non-formal institutions, and better communications are the key factor for assessing vulnerability and formulating coping strategies. This component is composed of six indices and 15 indicators. Except for the natural calamities index, all the indices showed significant difference in tribal and non tribal blocks. In totality, the components of weather perception and calamities, showed significant difference between tribal and non tribal blocks. Further, The vulnerability score of The tribal blocks is slightly higher due to higher dependency on natural resources and comparatively higher exposure.

1.5.1.10. Natural Resources

Natural resources are of paramount importance, not only from the economic point of view but also from the ecological point of view. It has a multifarious role in maintaining the ecological balance and moderation of the environment besides providing livelihood opportunities to people depending on forests. In the Malwa region, the livelihood of a large section of the population depends on forests. About 80% of the selected households collects fodder and fuelwood from the forest. This takes into account, distance from forest, time spent for fodder and fuel-wood collection, the quality of natural resources and their coping strategies all of which assessed and addressed. Six indexes and ten indicators were selected for assessing natural resource vulnerability of the Malwa region. The vulnerability score is slightly higher in the tribal regions. The higher vulnerability score might be attributed to higher rate of deforestation and lesser alternative of natural resources.

1.5.2. Vulnerability Assessment: IPCC Framework Approach

Further, the contribution of different dimensions (i.e., exposure, sensitivity and adaptive capacity) in the overall vulnerability of the Malwa region was assessed with the use of IPCC's vulnerability assessment approach. The overall vulnerability of the tribal blocks (Gandhwani – 0.693 and Bajna – 0.643) was higher as compared to non tribal blocks (Dhar - 0.453 and Jaora – 0.338). The higher vulnerability might be attributed to higher exposure and more sensitivity (poverty; family dependency; migration and geographical location) with lower adaptive capacity (assets, electricity and sanitation, irrigation, drinking water supply/facility).

1.5.3. Prioritization of Components through Dimension Reduction Technique

The data was collected on 80 indicators which are further reduced into ten components and it is believed that there is some redundancy in variables. This means some of the variables are correlated with one another, possibly because they are measuring the same construct. Therefore, to reduce the observed variables into a smaller number, the principal component analysis was used, that will account for most of the variance in the observed variables. Moreover, PCA would identify the most important factor (with the highest variance) that contributes significantly to sustainable livelihoods. The three factor solutions, which explained 76.27% of the variance, were preferred because of their previous theoretical support, and the 'leveling off' of eigen values on the scree plot. The most influential factor is agriculture with highest loadings followed by education, socio economic status and social network. The second component also consists of four factors i.e. food; water; natural resources; and weather perception and calamities. The third component is composed of two variables i.e. health and infrastructure.

1 Introduction

Climate change is recognized as a significant man-made global environmental challenge. International efforts to address climate change began with the adoption of the United Nations Framework Convention on Climate Change (UNFCCC). In its Article 1 UNFCCC defines climate change as "a change of climate which is attributed directly or indirectly to human activity and natural climate variability observed over comparable time periods". Climate change is a change in the statistical distribution of weather patterns when that change lasts for an extended period of time (i.e., decades to millions of years). Climate change may refer to a change in average weather conditions, or in the time variation of weather around longer-term average conditions.

Climate change is caused by a number of factors such as biotic processes, variations in solar radiation received by earth, plate tectonics, and volcanic eruptions. In addition, certain human activities have also been identified as significant causes of recent climate change. The human interference with the climate system is occurring, and climate change poses risks for human and natural systems (IPCC, 2014). The impact of climate change is particularly higher for some segments of society (poor and marginalized) and sectors (agriculture and forestry). Apart from this, it significantly affects health, safety and resource availability of poor and marginalized people (Cruz *et al.*, 2007 and Stern 2007), as they are confronted with limited livelihood options due to remoteness and less incentive to stay in balance with surrounding ecosystems. It has a multidimensional impact on humanity in terms of several socio-economic parameters like agriculture, human health, water scarcity, disease prevalence, etc. The probability of occurrence and magnitude of impacts are determined by weather and other climate related events (Brooks 2003; Goulden 2010). These interacting issues create complexities for assessing the vulnerability, which is a major concern, due to its guiding nature for future course of action on sustainability.

In India, around two-thirds (Census of India, 2011) of the population live in rural areas and are predominantly dependent on the climate sensitive sectors especially agriculture and forestry. Beyond providing the means of livelihood to a majority of the rural population, agriculture and forestry sectors also provide a raw material base for a number of industries like sugar, jute, textile, etc. Most of the rural households with agriculture based livelihood however, live on subsistence levels, enough only to meet their basic requirements. The subsistence level livelihood might be attributed to marginal and under-productive landholdings, periodic droughts, insecure land tenure and a reliance on seasonal agricultural and forest labour. The dependence on climate sensitive sectors makes them more vulnerable. Further, the ability of the rural poor to enhance their coping ability is constrained by a range of inter-related structural, social, economic and institutional barriers. Poverty, together with climatic extreme events, worsens the condition of the rural poor. It not only affects agricultural output and food security, but also escalates competition for resources including water, land and energy and trigger outbreaks of water and mosquito-borne diseases such as malaria. Moreover, the adaptive capacity of poor people especially farmers and forest dwellers is very low.

Projections indicate that the Indian mean annual temperature increased by 0.51 degree celsius per 100 years (1901 - 2007) (Kothawale *et al.,* 2010) and 0.2 degree celsius per decade (1971 – 2007). The decadal variation is much celsius in the increase in minimum temperature than maximum temperature. The mean maximum and minimum temperature has also significantly increased by 0.71 and 0.27 degree celsius per 100 years respectively. On the other side, Indian monsoon celsius well defined epochal variability with each epoch of approximately three decades. Additionally the frequency of hot days shows a gradual increasing trend and frequency of cold days shows a significant decreasing trend during the pre-monsoon season over the period 1970 – 2005 (INCCA, 2010). A number of studies have stated that India is likely to be hit hard by climate change. It is already one of the most disaster-prone nations in the world and about 1.2 billion people live in areas vulnerable to hazards such as floods, cyclones and droughts. Paroda and Kumar (2000) estimated that, by 2020, food grain demand will be almost 30-50% more than the demand during 2000 AD. Indian agriculture is facing challenges from several factors such as increased competition for land, water and labor from non-agricultural sectors and increasing climatic variability. There is a possibility of the loss of about 4-5 million tons in wheat production with every rise of 1 degree celsius temperature throughout the growing period with current land use (Aggarwal, 2008). Losses were also very significant in other crops, such as mustard, peas, tomatoes, onion, garlic and other vegetables, as well as fruit crops (Samra and Singh, 2004). In its National Communication Report to the UNFCCC, Indian Government mentioned that climate change is likely to impact all natural ecosystems as well as socio-economic systems. Along with this, India is among the countries that are most vulnerable to the adverse impact of climate change with economic losses of potentially 5 to 9 % of GDP (Stern, 2007).

Madhya Pradesh is the second largest state in the country in terms of area and endowed with diverse forest resources, and a rich history and heritage. The state has clocked 24.99% agricultural growth and was bestowed the prestigious Krishi Karman Award for its outstanding performance in the agriculture sector. As per the advance estimate figures of the Central Statistics Office (CSO), Madhya Pradesh registered 11.08 GSDP growth at constant prices in the year 2013-14. The state is also marked with a complex social structure, predominantly agrarian economy, and a difficult and inaccessible undulating terrain, along with scattered settlements over a vast area, which has been largely cutoff from mainstream development. This makes Madhya Pradesh one of the least developed states of India, with Human Development Index (HDI) value of 0.3753, which is below the national average of 0.467 (IHDR,2011). The state is home to a large tribal population with about 36 % population of the state belonging to the marginalized scheduled castes and scheduled tribes (20.27%) (Census of India, 2011) and 48.6% of the population estimated as living Below Poverty Line (BPL), with rural poverty ratio (53.6 %) exceeding the urban poverty ratio (35.1 %) significantly (Tendulkar Committee Report, 2009). Many areas in the state are vulnerable with considerable poverty and needs to scale up its anti-poverty measures. In addition the climate of the state is likely to change significantly and it has the potential to undermine existing efforts to tackle the mountain of poverty removal and the complex social development problems faced by the state. The projected increase in temperature in the 2030s for the state is around 1.8-2.0°C for A1B Special Report on Emissions Scenarios (SRES). By the 2080s, the maximum temperature is projected to rise between 3.4°C to 4.4°C with the northern region experiencing warmer temperatures. Further, the projections of rainfall for the period 2021 to 2050 indicates that there is likely to be a decrease in winter rainfall. Except for a few districts pre-monsoon and post monsoon rain are decreasing with no change or little increase in rainfall in most other parts of the state (MP-SAPCC, 2013).

In India, agricultural use comprises 80 to 90% of human water extraction, with almost 60% of this from groundwater. This large demand is part of an increasing trend with groundwater withdrawals increasing by 113-fold between 1950 and 1985. However, gross irrigated areas expanded only by four-fold, from 23 million to 90 million hectares for the period of 1951 to 1997 (Douglas *et al.,* 2006). As a result, mean ground water declined at a rate of 4.0±1.0 cm/year (Rodell *et al.,* 2009), as compared to declining rate of 20 cm/yr observed in 15 other states. In most of these states, the ground water resource is projected to dry out as soon as 2025 (Douglas *et al.,* 2009; Jha, 2001). Irrigation and drinking water supplies are thus likely to be a gigantic issue if water deficits are not managed properly.

Several studies have explicated the seriousness of recurring droughts in Madhya Pradesh with a far-reaching and deep impact on local livelihoods. Drought, lack of investment to produce year round cropping, degrading forest resources are all problems that will be exacerbated by climate change and make the task of delivering better livelihoods more difficult (Hedger *et al.,* 2010). Further, some communities of the developing world have special capabilities to counter climate change. The expected impact of climate change however, will exceed this capacity and continuously push these communities beyond their coping range. In order to strengthen the adaptive capacity of rural people, there is a need to understand the local vulnerability of the poor.

As far as vulnerability is concerned, human society is vulnerable to various causes such as disaster management, health, poverty, food security and climate change (Fussel 2007). The IPCC (2014) defines vulnerability as the "the propensity or predisposition to be adversely affected, sensitivity or susceptibility to harm and lack of capacity to cope and adapt" to climate stress; it may be represented as "either long-term changes in climate conditions, or by changes in climate variability, including the magnitude and frequency of extreme events". The social perspective of vulnerability is described as the "degree to which a system is susceptible to injury, damage or harm" (IPCC 2001), while the natural perspective of vulnerability addresses "a function of the character, magnitude and rate of climate variation to which a system is exposed, its sensitivity and its adaptive capacity (Goulden 2010)". Exposure, sensitivity, and adaptive capacity are inherently linked, and, therefore, can be hard to decipher (Gallopin 2006). Exposure is considered the potential impacts of climate variability and change; sensitivity is the degree to which the rural communities are affected by the exposure; and adaptive capacity is the system's ability to adapt, withstand or recover from the effects of the exposure (Ebi *et al.,* 2006; Hahn *et al.,* 2009). Exposure and sensitivity, almost inseparable properties of a system, are dependent on the interaction between the characteristics of the system and the climate stimulus attributes. The exposure and sensitivity of a system (e.g., community) to an environmental change risk (e.g., drought) reflect the likelihood of the system experiencing the particular condition and the occupancy and livelihood characteristics of the system which influence its sensitivity to such exposure (Smit and Wandel 2006). Adaptive capacity is similar to adaptability, coping ability, management capacity, stability, robustness, flexibility and resilience (Smithers *et al.,* 1997). These are potential agents for assessing the vulnerability due to climate change on the system.

There are broadly two interpretations of vulnerability – as an end point or as a starting point (Kelly and Adger, 2000; Brooks 2003; O'Brien *et al.,* 2004). Furthermore, the internal and external dimensions of vulnerability (Adger, 1996) and the concept of double structure of vulnerability (Bohle, 2001) are also significant in analyzing vulnerability. Vulnerability assessments are sometimes used as indicative tools for monitoring and evaluating of adaptation processes. Evaluations of these impacts aim to identify measures for strengthening adaptive capacity. One of the evaluation criteria revolves around the development of vulnerability indices. These indices seek to provide relative vulnerability scores for countries, regions or communities with the help of indicators, and they also identify adaptation strategies that are feasible and practical in communities (Smit and Wandel, 2006). The approach uses a specific set or combination of indicators, which is further aggregated into components or dimensions and measure vulnerability by computing indices. Components are case specific and may be qualitative characteristics of the system with the potentiality expressed as quantitative factors through the use of proxies (e.g., national or sub-national statistics) (Deems 2010). The components of vulnerability to climate change for a community revolves around the livelihood activity, and includes crop production and sufficiency; irrigation facility; farm size; crop variety; natural resources; livelihood strategies; education; social network; health and water. The development of components is crucial to a robust assessment of vulnerability and considerable attention must be attached to identifying representative proxies. The conversion of a system's characteristics into indicators facilitates the calculations of vulnerability in a transparent way that can be more easily defined and explained to the scientific communities and policy makers (Deems 2010).

The current study intended to understand the impact of climate change by evaluating the vulnerability of tribal and non tribal communities in the Malwa region situated in a relatively remote location. The vulnerability in the Malwa region is high and efforts at adaptation and mitigation are comparatively weak. The region is known as a pocket of poverty and the problem of poverty are further aggravated by limited livelihood opportunities, lower literacy, and scarcity of water and food. The vulnerability of the Malwa region is assessed by an indicator based approach so that the vulnerable sectors could be identified easily. Further, it would be helpful for planners/policymakers in formulating adaptation programmes for strengthening individual/collective assets. The study was undertaken with the following objectives.

- To identify the components of vulnerability and their impacts on livelihood of rural communities.
- To understand the perception of rural communities towards climate change, identify the primary causes and their coping mechanisms.
- To identify the vulnerable social groups and evaluate community-based coping or adaptation mechanisms to the impacts of change.
- To formulate recommendations (for planners and policymakers) to improve individual and collective assets, with the aim of enhancing the adaptive capacity and resilience of vulnerable communities in their responses to climatic change.

2 Literature Review

2.1. Climate Change and Vulnerability

Climate change has adversely impacted mankind and extreme events are unevenly distributed between nations, regions and communities. Climate change will lead to changes in geophysical, biological and socio-economic systems. An impact describes a specific change in a system caused by its exposure to climate change. Further, the impact will worsen the existing developmental challenges, undermining the adaptation capacity and increasing vulnerability of the people.

The scientific use of 'vulnerability' has its roots in geography and natural hazards research, however, this term is now a central concept in a variety of research contexts such as natural hazards and disaster management; ecology; public health; poverty and development; secure livelihoods and famine; sustainability science; land change, and climate impacts and adaptation. A simple, elegant definition is often used in literature and arises from its Latin origins. The root of vulnerable is "*vulnerare*" meaning "to wound"; therefore, vulnerability can be basically described as: "*the capacity to be wounded*" (Kates *et al.*, 1985, in Dow, 1992).

Amartya Sen (1981) proposed that vulnerability could be explained by a lack of entitlements to things like food security, sustainable livelihoods, social structures, etc., through institutional, political, technological and other constraints. The Intergovernmental Panel on Climate Change (IPCC), in its Second Assessment Report, defines vulnerability as "the extent to which climate change may damage or harm a system". It adds that vulnerability "depends not only on a system's sensitivity, but also on its ability to adapt to new climatic conditions" (Watson *et al.*, 1996). Further, the location and context specificity of vulnerability with uncertainty is discussed deliberately by Handmer in 1999. He emphasized that climate vulnerabilities are location and context specific, with climate uncertainty increasing as analytical resolution in terms of space and time increases. This uncertainty must be tackled by bold policy interventions that address the social and economic factors that drive climatic vulnerability; while taking into account human behavior, culture and institutional capacity of particular communities, alongside the biophysical impacts of a changing climate. At the Sixth Conference of the Parties to the UNFCCC (COP-6), Watson defines vulnerability as the extent to which a natural or social system is susceptible to sustaining damage from climate change. It is a function of the magnitude of climate change, the sensitivity of the system to the change in climate and the ability to adapt the system to climate change. Hence, a highly vulnerable system is one that is highly sensitive to modest changes in climate and one for which the ability to adapt is severely constrained (IPCC 2000).

The IPCC report, "The Regional Impacts of Climate Change: An Assessment of Vulnerability" argues that the vulnerability of a region depends to a great extent on its wealth, and that poverty limits adaptive capabilities. According to the Second Assessment Report, vulnerability depends on the level of economic development and institutions. The report argues that socio-economic systems "typically are more vulnerable in developing countries where economic and institutional circumstances are less favorable". The report continues that vulnerability is highest where there is "the greatest sensitivity to climate change and the least adaptability". Along with this, a number of researches suggest that the social dimensions of the social-ecological system are the most important factors in aggravating vulnerability, and also in reducing it, through improved resilience and adaptation measures. Since 1980, a variety of definitions of vulnerability has been proposed. It is generally taken to be the ability to anticipate, resist, cope with, and respond to a hazard (Blaikie *et al.,* 1994) and it occurs due to various causes such as disaster management, public health, poverty; food security, ecology and climate change (Fussel 2007).

The social perspective of vulnerability was defined by the IPCC (2001) as "degree to which a system is susceptible to or unable to cope with, adverse effects of climate change, including climate variability, change and extremes". On the other side, the natural perspective of vulnerability is addressed as "a function of the character, magnitude and rate of climate variation to which a system is exposed, its sensitivity and its adaptive capacity" (McCarthy *et al.,*. 2001; Goulden 2010). Exposure, sensitivity, and adaptive capacity are inherently linked and therefore, can be hard to decipher (Gallopin 2006). Exposure is considered the potential impacts of climate variability and change; sensitivity is the degree to which rural communities are affected by the exposure; and adaptive capacity is the system's ability to adapt, withstand or recover from the effects of the exposure (Ebi *et al.,* 2006). Further, vulnerability is defined as a measure of the degree of defenselessness or propensity of a system (a community) to be damaged or harmed from the effects of climate change. The IPCC defines vulnerability as "the propensity or predisposition to be adversely affected". Vulnerability encompasses "a variety of elements including sensitivity or susceptibility to harm and lack of capacity to cope and adapt" (IPCC 2013).

Vulnerability to climate change is determined by the physical impacts of climate change on the availability of resources that people depend on, and the accessibility of people to scarce resources. In this light, vulnerability can be understood as a social construction which is influenced by institutional, economic and biophysical features (Adger *et al.,* 2003). Vulnerability is not strictly synonymous with poverty, although poverty and marginalization frequently are drivers of vulnerability and the inability to cope and adapt; likewise, vulnerability to climate change is likely to exacerbate existing vulnerabilities and create new vulnerabilities.

2.2. Vulnerability and Adaptive Capacity

Earlier, social vulnerability was largely underemphasized in the assessment of the impacts of climate change and climate extremes. Therefore, in 1999, Adger designed a framework for analyzing social vulnerability. He defines vulnerability as the exposure of individuals or collective groups to livelihood stress as a result of the impacts of such environmental change. It constitutes individual and collective aspects, which can be disaggregated, but are linked through the political economy of markets and institutions.

After analyzing the social vulnerability which was largely underemphasized, Adger *et al.,* (2000) provide an approach by which vulnerability and adaptive capacity are assessed. The first step is to identify the climate change related risks in any given area, as well as other socio-economic and environmental stressors. The second step is to identify sensitivities in terms of a limited capacity to cope and respond to environmental and social changes. Step three is to assess the existing capacity of the given community to current impacts, and the extent to which adaptation activities are already supported by existing policy and support programs. The fourth step is then to assess the ability of any given community to adapt to the future impacts of climate change given the results of the previous steps and, finally, recommendations for policy or strategy are suggested. In 2001, Moss designed an index on vulnerability resilience to climate change. He identified ten proxies for five climate sensitive sectors, and seven proxies for three sectors of coping and adaptive capacity. The proxies are aggregated and finalized into a vulnerability-resilience index. Again, in 2003, Adger reviewed the emerging perspectives on collective action and social capital. He argued that social capital is increasingly understood within economics to have public and private elements, both of which are based on trust, reputation, and reciprocal action. The public-good aspects of particular forms of social capital are pertinent elements of adaptive capacity in interacting with natural capital, and in relation to the performance of institutions that cope with the risks of changes in climate.

Additionally, the associated risk of climate change to society may exacerbate ongoing social and economic challenges, particularly for those dependent on climate sensitive resources such as agriculture, fisheries and livelihood. Adger (2003) explored the nature of risk and vulnerability in the context of climate change and reviewed the evidence on present-day adaptation in developing countries and on coordinated international action on future adaptation. He also added that all societies are fundamentally adaptive and emphasized on the need for enhancing adaptive capacity to face both present and future climate change outside their experienced coping range.

Further, the simplification of the complex relationship of vulnerability, adaptation, risk applicable to climate variability and change in human systems are presented in a conceptual framework by Brooks (2003). He also identified the somewhat complex terminology and confusion surrounding vulnerability assessments. In this study, Brooks attempts to resolve conflict between social and biological assessments by highlighting their differences, and analysing the definition of risk in respect of hazards and the IPCC's definition of vulnerability. An examination of hazards is then given before the different conceptual timescales of vulnerability are expanded to give a platform for quantifying adaptive capacity. He suggested a slight change in terminology, adding "adaptive likelihood", rather than capacity, and places social vulnerability in the risk management framework by accepting the similarity of risk between the biophysical and natural hazard concepts. The measurement of risk in social vulnerability is, therefore, suggested to be either in terms of the mortality of human, or in factors such as educational status, coping ability or economic well-being.

The role of adaptive capacity in vulnerability assessment is interpreted in an "end point and starting point" approach. The first is the "end point" approach, which sees vulnerability as being determined by adaptive capacity and something that resides in the future. The second view of vulnerability is as the "starting point", which shows vulnerability as already being present and determining adaptive capacity. O'Brien emphasized that the differing interpretations of vulnerability that reside across the multidisciplinary scientific and policy community have significant implications for how the effects of climate change can be diagnosed and resolved. He concluded that viewing vulnerability as an end point does not address the underlying causes of vulnerability, but instead focuses upon the symptoms and therefore, greater insight can be found by analysing current and past contexts, by seeing vulnerability as the starting point (O'Brien, 2004).

Earlier to O'Brien, a multi scaled, integrated framework for assessing vulnerabilities and adaptive capacity was developed by Dolan and Walker in 2003. The determinants of adaptive capacity included access and distribution of wealth; risk perception; social capital and community structure and institutional frameworks that address climate change hazards. These are identified at the individual and community level and situated within larger regional, national and international settings. The social-ecological system (SES) that determines future trajectories of a system are: resilience, adaptability and transformability (Walker *et al.,* 2004). The implication of interpretation of SES dynamics for sustainability science include changing the focus from seeking optimal states and the determinants of maximum sustainable yield, to resilience analysis, adaptive resource management, and adaptive governance. He concluded that the resilience of a system needs to be considered in terms of the attributes that govern the system's dynamics.

In 2005, Adger emphasized that the impact and response of climate change in physical and ecological systems, as well as in human adjustments relate to resource availability and risk at different spatial and societal scales. Along with this, he outlines a set of normative evaluative criteria for judging the success of adaptations at different scales that involve new and challenging institutional processes. The concept of adaptation of human communities to global changes, especially climate change, in the context of adaptive capacity and vulnerability, is reviewed by Smit *et al.*, (2006). It focuses on scholarship that contributes to practical implementation of adaptation on the community scale. They recommended that climate is considered together with other environmental and social stresses, and adaptations are mostly integrated or mainstreamed into other resource management, disaster preparedness and sustainable development programs. After the concept of practical implementation of adaptation program at the grass root level, Fussel (2007) formulated the concept of planned adaptation to climate change. It denotes actions undertaken to reduce the risks and capitalize on the opportunities associated with global climate change. The planned adaptation starts with an explanation of key adaptation concepts, a description of the diversity of adaptation contexts, and a discussion of key prerequisites for effective adaptation. Moreover, Thomas (2007) emphasized the complexity in coping and adaptation, and the factors that influence the decision. He identified the range of specific coping and adaptation strategies that are employed by farmers to respond to climate shifts, some generic across regions, and some facilitated by specific local factors.

The complexity and multitude of vulnerability approaches and frameworks was highlighted in 2007 by Fussel. He attempts to bring clarity by presenting a generally applicable conceptual framework for vulnerability that incorporates classification of vulnerable situations and concepts based on four dimensions of vulnerability. These are: (1) the system, such as the population, geographic region or human environment; (2) the attribution of concern, indicating the valued elements that are under threat from exposure to a hazard; (3) the hazard; and (4) the temporal reference, being the time or time period of interest. A method for allocating and classifying a description of a vulnerability assessment is also given, whether it is in the internal or external sphere of socio-economic or biophysical domain. The framework presented is argued to enable scholars and those involved in vulnerability assessments to see how and why vulnerability concepts differ, their diversity, and their clarity to identify interpretations of vulnerability in the multitude of available assessments. A review of past conceptual frameworks at the initial stages of this paper brings relevance, depth and reference for the author's framework, assisting in the identification of gaps between the varieties of interdisciplinary vulnerability approaches.

Adaptation to climate change is controlled by a number of factors and it also varies according to geographic location and the local adaptation mechanism. The factors that affect their choice of adaptation and the barriers to adaptation in the Nile Basin of Ethiopia were studied by Deressa *et al.* in 2008. A large number of methods of farmer's adaptation to climate change were identified in the study. Such as the use of different crop varieties, tree planting, soil conservation, early and late planting, and irrigation. He suggested that the level of education, gender, age, and wealth of the head of household; access to extension and credit; information on climate, social capital, agro ecological settings, and temperature influence farmers' choices. He emphasized that the main barriers include lack of information on adaptation methods and financial constraints. Moreover, the analysis reveals that the age of the household head, wealth, information on climate change, social capital, and agro ecological settings have significant effects on farmers' perceptions of climate change.

The vulnerability and adaptive capacity assessment of the water stressed region of Lakhwar watershed in Uttarakhand State, India, is carried by Kelkar *et al.*, 2008. He identified highly water stressed micro-watersheds by modeling using variables such as surface runoff, soil moisture development, lateral runoff, and groundwater recharge. The results were shared with communities in two villages, and timeline exercises were carried out to allow them to trace past developments that have impacted their lives and livelihoods, and stimulate discussion about future changes and possible adaptation interventions. He suggested that the current coping capacity of people in the region to climate variability and water stress is quite low. Households are considerably dependent on low-value rain fed agriculture. Institutional capacity is also poor, particularly in terms of connectivity and the availability of formal credit, which constrains their ability to use their agricultural skills and assets more effectively. The communities have limited skilled human resources, which limits their options in seeking off-farm employment opportunities.

Awareness and socio-economic status of farmers are determining factors for designing and implementing adaption programmes. The awareness of farmers on climate change and variability, along with their socio-economic status was discussed by Mertz *et al.,* (2008). He recorded the perception on climate change and the strategies for coping and adaptation by sedentary farmers in the savanna zone of central Senegal. He found that households were aware of climate variability and identify wind and occasional excess rainfall as the most destructive climate factors. Poor livestock health, however, reduced crop yields and a range of climatic factors, hindering adaptive capacity and augmenting vulnerability. Further, he emphasized that change in land use and livelihood strategies is driven by adaptation to a range of factors.

The relationship with climate induced shocks and communities' adaptation mechanism was analyzed by McSweeney (2010). He comes up with compelling evidence that communities can seize on the window of opportunity created by climate-induced shocks to generate sustained social-ecological improvement. He suggested that future interventions should foster local capacities for endogenous institutional change to enhance community resilience to climate shocks. The longitudinal analysis of the study reveals that disaster enabled the poor to initiate an institutional change that led to more equitable land distribution, slowed primary forest conversion, and positioned the community well to cope with comparable flooding occurring 10 year later.

The integrated climate information system in risk management and adaptation planning is of critical importance and discussed deliberately by Munang *et al.,* (2010). He added that the provision of climate information and an understanding of ecosystem responses to climate change and variability urgently need to underpin any planning for the future. Moreover, a substantial mindset shift to fully recognize the fundamental role of ecosystems as life-supporting systems is urgently needed. The value given to ecosystems and the magnitude of effort to manage them has to be based on this mere fact and, indeed, it should be an integral part of any climate change agreement.

The impact of climate change is expected to be higher for climate sensitive sectors, i.e., agriculture, forestry and fisheries. The impact of climate change on agricultural productivity, food security and prices, water availability and human health is higher, especially in developing countries. Along with this, the capacity in these countries and continents to adapt is very low. Ozor *et al.,* (2012) conducted a study on the same in Africa and his result showed that the majority of respondents (84%) were aware that some climate change characteristics such as uncertainties at the onset of the farming season, extreme weather events including flooding and droughts, pests, diseases, weed infestation, and land degradation have all been on the increase. The most significant effects of climate change that manifested in the area were declining soil fertility and weed infestation. Some of the adaptation strategies adopted by farmers include increased weeding, changing the timing of farm operations, and processing of crops to reduce post-harvest losses. Although a majority of the respondents were aware of government policies aimed at protecting the environment, most of them agreed that these policies were not being effectively implemented. A mutually inclusive framework, comprising both indigenous and modern techniques, processes, practices and technologies, was then developed from the study in order to guide farmers in adapting to climate change effects/impacts.

2.3. Vulnerability Indicators and Indices

In the last few decades, a number of Indices relating to vulnerability, sustainability, and quality of life has been developed with different sets of parameters such as Human Development Index (HDI) emphasizing on a standard measure of human development in the country. The HDI incorporates life expectancy, health, education, and standard of living indicators for an overall picture of national well being (UNDP 1990; 2005; Burd-Sharps *et al.,* 2008, 2009). While the number of variables is small, the goal of the HDI is to compare the relative levels of development and improvements in well-being among nations, and is the gold standard for assessing the human condition at the country level.

The alternative of HDI, developed by Prescott-Allen in 2001 is called the Human Well-being Index (HWI). The HWI analyses demographic data along with measures of community (political rights, crime, internet users, and peace and order), and social equity (gender and income). The HWI uses a much larger pool of variables (33) in which to derive their five indicators or sub-indices for 180 countries. The five sub-indices represent health and population, wealth, community, equity, and knowledge.

The Environmental Vulnerability Index (EVI) explicitly describes the factors that negatively influence the sustainable development process. The approach, developed by the South Pacific Applied Geoscience Commission

(SOPAC), was one of the earliest efforts and examines vulnerability to environmental change (Kaly *et al.,* 1999 and SOPAC 2005). The scale of analysis is at the country level with the emphasis on Small Island Developing States (SIDS). The EVI excludes human systems, and so is more like an exposure assessment rather than an integrated vulnerability assessment. The EVI uses 50 biophysical or natural environment indicators grouped into three sub-indices (hazards; resistance; damage) and then applies these to the following policy relevant issues or threats: climate change, biodiversity, water, agriculture and fisheries, human health, desertification, and exposure to natural disasters. Like the EVI, the Environmental Sustainability Index (ESI) is a composite index consisting of 76 variables reduced to 21 sub-indices to create an overall sustainability score by summing each sub-index and then taking the average. It assesses the sustainability of 146 nations based on five core components: environmental systems, environmental stresses, human vulnerability to environmental stresses, social and institutional capacity, and global stewardship (Esty *et al.,* 2005). Country rankings provide useful information on the relative level of sustainability, but an examination of the sub-indices (especially over a number of years) provides valuable policy information on improvements in meeting sustainability goals.

The Prevalent Vulnerability Index (PVI), designed by Cardona (2005), consists of 24 indicators. The choice of indicators imply that the PVI is largely rooted in the political-ecological tradition of vulnerability science and focuses on social, economic, institutional, and infrastructural capacity to recover from natural hazards, or the lack thereof. The PVI employs Min-Max normalization to restrict the value range of indicators between zero and one. The PVI's indicators are aggregated into three sub-indices with varying weights, and then summed to generate the final index score. The PVI for Latin America employs vulnerability-relevant indicators, but suffers from methodological weaknesses when it comes to indicator aggregation and weighting (Gall 2007).

A theory-driven aggregate index of social vulnerability was formed through the weighted average of five composite sub-indices: economic well-being and stability (20%), demographic structure (20%), institutional stability and strength of public infrastructure (40%), global interconnectivity (10%), and dependence on natural resources (10%) (Vicent, 2004). The aim of the research is to create an index to empirically assess relative levels of social vulnerability to climate change-induced variations in water availability and allow cross-country comparison in Africa. Moreover, Adger (2004) focuses upon the development of new vulnerability and adaptive capacity indicators through a comprehensive analysis of vulnerability and the creation of a working database for socioeconomic indicators to vulnerability from climate change. Through the use of literature reviews, practitioner meetings and discussions with key individuals, a conceptual framework was developed with indicators for vulnerability and adaptations. The development of risk indicators was then completed with the use of the Emergency Events Database (EM-DAT), enabling a useful outcome method with which to check the validity of proxies. Predictive indicators were then developed along with proxy variables of vulnerability before the identification of indicators began. He found the key areas of health and social capital as being inadequately represented in previous vulnerability studies and, along with indicators of education and governance, they highlighted their usefulness in assessing vulnerability to climate hazards.

The national level indicator for measuring climate change vulnerability was developed by Eriksen in 2006. He compared the indicators of five national-level studies: the vulnerability–resilience indicators by Moss *et al.*, (2001); the Environmental Sustainability Index from the World Economic Forum (2002); dimensions of vulnerability by Downing *et al.*, (1995); the Index of Human Insecurity (IHI) from Longergan *et al.*, (1999); and the country-level risk measures by Brooks and Adger (2003). The studies were then categorized by a number of indicators and countries, and their incorporation of environmental resources, economic resources, social conditions and institutional capacity. The study concludes that national-level indicator studies have thus far failed to encompass adequately the processes of vulnerability, and thus greater emphasis on process-based approaches is suggested. The authors make three specific recommendations to develop credible indicators: the differentiation of vulnerability at all levels: increasing the transparency of authors' assumptions and premises of vulnerability; and the importance of verifying findings.

Vulnerability and capacity of the communities change constantly, reflecting prevailing social, economic, cultural and political circumstances. This can also dramatically intensify the effects of a disaster. To respond more effectively to people's local concerns – be it unsafe drinking water, exposure to malaria or traffic accidents, the International Federation of Red Cross and Red Crescent Societies (2006) designed a straight forward, easier to understand

and more useful guide named Vulnerability and Capacity Assessment Tool, for practitioners. The VCA is concerned with collecting, analyzing and systematizing information on a given community's vulnerability to hazards in a structured and meaningful way. This information is then used to diagnose the key risks and existing capacities of the community, ultimately leading to activities aimed at reducing people's vulnerability to potential disasters and increasing their capacity to survive them and resume their lives.

Vicent (2007) uses two empirical adaptive capacity indices for use at national and household scales. He described a national index for cross-country comparison in Africa and a household index for cross-household comparison in a village in Limpopo province, South Africa. He emphasized that indicators and indices are common tools in adaptive capacity assessment, but their construction embodies many uncertainties, not the least of which is their scale specificity. The decisions made at each stage of construction illuminates the degree of uncertainty involved when assessing adaptive capacity, and how this uncertainty is compounded when looking across different scales of analysis.

A study by Ebert in 2008 specifically focused upon measuring social vulnerability in a neighborhood of Honduras through an alternative method of using high-resolution images to gain proxy variables, onto which object-oriented analysis (OOA) was conducted for further definition. Along with these data, elevation and hazards were also included and, in total, 47 proxy variables were identified, including socioeconomic status of households and neighborhood characteristics. An SVI was formed using existing data from the area combined with Geographic Information System (GIS) data. The study concluded that GIS data can be useful for the measurement and assessment of social vulnerability, especially when combined with further context specific data, producing a more cost-effective study with wider coverage. The benefits of this approach not with standing, the authors highlight that it is in no respect a comprehensive assessment of social vulnerability, and that integrating traditional field surveys along with GIS mapping would result in greater depth and assist disaster management in areas with scarce data.

In a joint effort, Emory University (U.S.A.) and Cooperative for Assistance and Relief Everywhere (CARE)-Mozambique, Hahn (2009) developed an

Livelihood Vulnerability Index (LVI) for estimating the climate change vulnerability of two districts of Mozambique: Mabote and Moma. The LVI combined seven components or indicators for vulnerability: livelihoods, socio demographics, social networks, health, natural disasters and climate variability, and food and water security. The various components of vulnerability were then weighted equally and combined into an overall LVI. An alternative approach was then formed by combining the LVI model and its seven components with the IPCC's three factors of vulnerability (exposure, sensitivity and adaptive capacity). Implemented at the district level, the authors highlight the relevance of this method for resource and adaptation planning, noting the applicability of the LVI for development organizations, practitioners and policy-makers to assess vulnerability. The authors highlight the limitations and benefits of this approach, with the former being the simplification of complex reality into only seven components (reducing the variance of results) and the latter being the usefulness in district-level planning and the ability for comparing vulnerability between locations.

The Climate Vulnerability and Capacity Analysis (CVCA) created by CARE International builds upon its community-based approach, with the belief that empowerment through participation and knowledge sharing can enable transformation. The CVCA uses participatory methods of analysis, aiming to link community knowledge with scientific data in order to create a mutually beneficial process for analysis and adaptation of vulnerability. The method recognizes the importance of context-specific analysis and increasing dialogue within communities at the regional and national levels, and it combines prior work on poverty and vulnerability with that of disaster risk reduction (DRR). The CVCA is structured along four sections in the community-based approach: resilient livelihoods, DRR, capacity development, and addressing the underlying causes of vulnerability. These help guide the vulnerability analysis applied to the national, local government/community and household/individual level.

Again, Vincent and Cull (2010) constructed a theoretically-derived index for identifying households with the highest relative vulnerability, target of adaptation interventions and to evaluate the success of such interventions in reducing vulnerability. The index was based on a sustainable livelihood framework and useful for assessing household level social vulnerability to climate change, in terms of multiple dimensions of vulnerability. The index is formed from the weighted aggregation of 5 composite sub-indices, themselves formed from one or more indicators. He exemplifies the use of the index in a rural dryland setting in South Africa, where changes in water availability is the climate hazard. Further, the climate adaptation strategies of farmers in the Limpopo Basin of South Africa are studied with multinomial logit analysis (Gbetibouo *et al.,* 2010). The adaptation responses suggest that access to water, credit, extension services, off-farm income, employment opportunities, tenure security, farmers' asset base and farming experience are key to enhancing farmers' adaptive capacity. He noticed that the lack of access to credit and water were cited as the main factors inhibiting adaptation. Common adaptation responses reported included diversifying crops, changing varieties and planting dates, using irrigation, and supplementing livestock feed. This implies that appropriate government interventions to improve farmers' access and data on the status of these factors are needed for reducing vulnerability of farmers to climate adversities.

Subsequent to Hahn's Livelihood Vulnerability Index (LVI) (2009), Urothody *et al.,* in 2010, designed the Livelihood Vulnerability Index (LVI) and the Livelihood Effect Index (LEI) in Lete and Kunjo Village Development Committees (VDCs) in Mustang District. The indexes are computed based on primary data collected from VDCs. Both indexes reflect the relative differences between the two VDCs in terms of vulnerability to climate change impacts and factors contributing to it. They recommended that both could, therefore, usefully form the basis for a nationally applicable index to identify and prioritize mitigation needs.

A unique review of the frameworks and models used in the Australian agriculture sector to assess vulnerability to climate change was designed in 2011. Initially, Pearson reviewed the various conceptual frameworks in the published literature, then identified the models currently being used to highlight specific gaps in need of further attention, with the use of an expert workshop of biophysical, economic and social scientists. Six criteria were used to review the vulnerability assessments: scope, scale, operational status, accessibility, validation and data availability. The review found four main types of research approaches in Australian agriculture, the first being biophysical productivity models that identify yield production of plants and animals. The second is bio-economic models, which focus on the economic productivity and incomes of farms to assess future changes. The next is a method for regional change planning that uses scenario assessments in various regions of agricultural industries, and, finally, there is an adaptive capacity model using the rural livelihood framework for agricultural industries. The review identifies the current vulnerability assessments in the agriculture sector as being mainly outcome-based, using predominantly biophysical models. Finally, Pearson concluded by strongly suggesting a shift in Australian vulnerability assessments to more contextual frameworks, providing more emphasis on accurate estimations of crops, infrastructure and socio-economic systems.

A conceptual framework for vulnerability indicators was developed by Hinkel (2011). He reviewed a range of arguments used in vulnerability indicators, such as the deductive, inductive, normative and non-substantial. Six types of problems that indicators of vulnerability should address are presented : (i) the identification of mitigation targets (ii) vulnerable people (iii) communities and regions (iv) raising awareness (v) allocating adaptation funds (vi) monitoring adaptation policy and conducting scientific research. The review found that only indicators identifying vulnerable people were appropriate, with those being used for the other five problems being deemed an inappropriate concept or method to use. He argues that measuring vulnerability is impossible and that a change of terminology is required within the field of vulnerability research. The importance of recognizing the difference between vulnerability and harm indicators is also noted, with the latter not including future dimensions.

The vulnerability and capacity assessment approach designed by the International Centre for Integrated Mountain Development (ICIMOD) in 2011 is a combination of the vulnerability assessment and sustainable livelihoods approaches (SLA). VCAs are carried out applying a gender perspective and paying attention to marginalized social groups (e.g., based on caste or ethnicity). The objective of the VCA approach is to identify not only the key vulnerabilities of mountain communities and the underlying causes of these vulnerabilities, but also the community's inherent livelihood assets and capacities, and then to suggest ways to strengthen these assets and capacities to enhance the community's resilience.

Recently, climate change vulnerability index for sustainable livelihoods was proposed by Singh and Nair (2013) to estimate the vulnerability of poor agro-pastoralists in the Bhilwara and Tonk districts of Rajasthan, India. They added that the existing studies of livelihood vulnerability assessments often do not capture the dynamics and functioning of interconnected systems. Therefore, they used a semi-quantitative fuzzy cognitive mapping approach to understand interconnected interactions occurring in the climate-human-environment interaction space. This approach enables the quantification of peoples' perceptions of perturbations to a given climate exposure and their adaptation mechanisms in relative terms. This was combined with a sustainable livelihood approach for an improved understanding of asset classes sensitive to climate change and assets serving as adaptive capacities.

The Poverty and Vulnerability Assessment (PVA) tool of ICIMOD is designed to capture key elements of poverty, vulnerability, and adaptive capacity in mountain contexts for the Hindu Kush Himalayan (HKH) region. The PVA combines general predictors of poverty with indicators that are particularly relevant in mountain contexts, where factors such as physical accessibility and a lack of access to basic facilities often exacerbate poverty and vulnerability to stresses such as those related to climate variability and change. The inclusion of indicators of adaptive capacity acknowledges that, while mountain households in the HKH region may experience poverty and vulnerability, the communities they represent are also repositories of experience and knowledge that can help people cope with stress and adapt to change. Further, it is particularly relevant in the context of development interventions aimed at building resilience and facilitating adaptation to climate variability and change (ICIMOD, 2014).

3 Study Area

3.1. Malwa Region

The study has been undertaken in Malwa region of Madhya Pradesh. The name Malwa is derived from the Sanskrit *malav* (abode of Laksmi). The Malwa Plateau is of volcanic origin and the region lies in north-central India (21°10'N 73°45'E and 25°10'N 79°14'E). It is bounded by the Madhya Bharat Plateau and Bundelkhand upland to the north, the Vindhya Range to the east and south, and the Gujarat Plains to the west. It comprises central Madhya Pradesh and southeastern Rajasthan. The region comprises Dewas, Dhar, Indore, Mandsaur, Neemuch, Rajgarh, Ratlam, Sajapur, Ujjain, part of Guna and Sehore of Madhya Pradesh; and Jhalawar and parts of Banswara and Pratapgarh of Rajasthan. The plateau is an extension of the Deccan Traps, formed between 60 and 68 million years ago at the end of the Cretaceous period with an average elevation of 500 m.

3.1.1. Demography

The Malwa region is dominantly inhabited by the Bhil tribe (Meos, Barelas and Patelias). In addition, Meenas, Dawoodi Bohras (Shia Muslims from Gujarat) and Patidars (Kurmis of Punjab) reside in the region with remarkable variation in dialects and social life. A significant number of Marwaris, Jats, Sindhis, Jains and Rajputs also live in the region. The region is home to smaller numbers of Goan Catholics, Anglo-Indians, Punjabis and Parsis or Zoroastrians.

The population of the Malwa region was about 15,685,878 in 2011. The annual birth rate in the region was 31.6 per 1000, and the death rate 10.3. The infant mortality rate was 93.8, slightly higher than the overall rate for the Madhya Pradesh state. The literacy in the region is 70.27%.

Further, the region has known pockets of poverty and is inhabited by a higher concentration of scheduled castes (13.86%). The poverty is further aggravated by limited livelihood opportunities, lesser literacy, subsistence agriculture, degradation of natural resources, scarcity of water (surface and ground), deprivation of special government programmes and the absence of a strong people's organization /civil society/NGO all major constraints for development.

3.1.2. Agriculture

The economy of the Malwa is dominated by the agricultural sector. The soil of the region is black volcanic, which is suitable for cultivation of wheat, cotton and soyabean. Overall, the main crops are soyabean, jowar, rice, wheat, coarse millet, opium, peanuts, pulses, cotton, linseed and sesame. The region has been one of the important producers of opium in the world. The districts of Malwa region has acute water shortage and suffers degradation of soil and biomass at an unprecedented scale. The loss of biomass and undulating topography accelerates topsoil erosion and hence productivity of agricultural land is decreasing. Together with this, the agriculture of the region is further constrained by a large proportion of small and marginal farmers.

3.1.3. Livestock

The majority of the households in Malwa region possess one or two livestock, preferably cow and buffalo. Apart from this, ox, goat, chicken and sheep are also reared on a large scale. Livestock in the region is not the major source of income, except in a few villages. The inaccessibility to hybrid livestock, infrastructural constraints and remoteness from main consumer markets pose severe challenges for perishable products. In the region livestock act as "savings" and as a buffer against shocks and stresses, whether in the farm or in the domestic context.

3.1.4. Forest

Forest cover in Malwa region is very less, i.e., 09.21%, but the economic importance of forest is somewhat good due to patches of teak. The main tree species are Butea, Bombax, Snogeissus, Acacia and Boswellia. Forest in the region plays a vital role in the livelihood of villagers in fulfilling their requirements for fodder and fuelwood. Fodder and fuelwood which is in great demand, however, is not sufficient in supply to meet the local demand, which is, perhaps, one reason for the higher rate of forest degradation

3.1.5. River

The plateau generally slopes towards the north. The western part of the region is drained by the Mahi River, while the Chambal River drains the central part, and the Betwa River and the headwaters of the Dhasan and Ken Rivers drain the east. The Shipra River is of historical importance because of the Simhasth mela, held every 12 years. Other notable rivers are Parbati, Gambhir and Choti Kali Sindh. The main purpose of rivers in the region is irrigation and water availability to livestock. Rivers however, do not provide sufficient irrigation throughout the year; therefore, groundwater is overexploited and the situation is considered critical

3.1.6. Climate

The year is popularly divided into three seasons: summer, the rains, and winter. Summer extends over the months of Chaitra to Jyestha (mid-March to mid-May). The average maximum temperature during the summer months is 40 °C, which typically rises to around 45 °C on a few days. The rainy season starts with the first showers of Ashaadha (mid-June) and extends to the middle of Ashvin (September). Most of the rain falls during the southwest monsoon spell, and ranges from about 80 cm in the west to about 10.5 cm in the east. Winter is the longest of the three seasons, extending for about five months, i.e., mid-Ashvin to Phalgun (October to mid-March). The average daily minimum temperature ranges from 6 °C to 9 °C, though on some nights, it can fall as low as 3 °C. Some cultivators believe that an occasional winter shower during the months of Pausha and Maagha, known as Mawta, is helpful to the early summer wheat and germ crops.

3.2. District Dhar: *An overview*

The district of Dhar is located in the south western part of the State of Madhya Pradesh. It is located between latitude 22^0 N - $22^049'$ N and longitude - $75^006'$- $75^042'$ E. The District is Located 266 KM east towards the state capital, Bhopal. The average altitude of Dhar District is 588 meters above sea level. The district extends over three physiographic divisions,

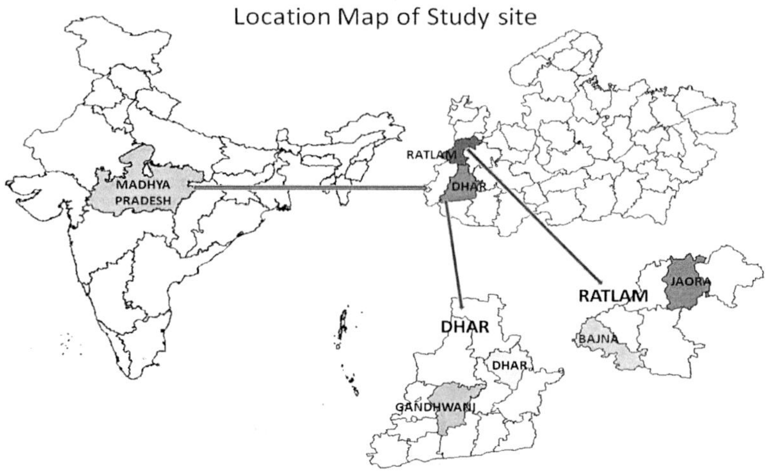

namely, the Malwa in the north, the Vindhyachal range in the central region and the Narmada Valley along the southern boundary. The climate of the district is subtropical with three distinct seasons, viz., winter (December to February), summer (March to May), and rainy (June to October). The summer temperature varies between 31°C to 43°C and April and May are the hottest months. The average precipitation of the district is 824 mm and most of the precipitation falls in July and August. The Total population of the district is 2184672, wherein 1771557 people are living in rural areas. It is Seventh largest district in the state by population. The district is characterized by high tribal population (54.50 %), majorly rain-fed agriculture, lesser number of rainy days and low annual rainfall (774 mm annual average rainfall over a period of 50 years), critical ground water (average 31 ft depth of water table over five years), undulating terrain, low vegetation cover (09 %) and a large proportion of wasteland (12.51 %), making it an ideal site for assessing vulnerability and capacity of the region

3.2.1. Selection of Blocks

Dhar and Gandhwani blocks of Dhar district were selected for the study. The total population of these blocks is 578423 (male-301567 and female-276856) in Dhar and 156046 (Male-77672 and Female-78374) in Gandhwani. The SC/ST population of Dhar is 09.25/36.44% and Gandhawani is 01.47/90.74%. The literacy of the Gandhwani block (34.83%) is much lower than the average literacy of the district, i.e., 47.16%. The literacy of Dhar block (57.18%), however, is higher than the average literacy of the district. The prime profession of the villagers is agriculture and average landholding in the block is 0.21 hectare. The Gandhwani block is characterized by rain-fed agriculture, critical ground water level, undulating terrain, low vegetation cover and large proportion of wasteland. The agriculture of Dhar block is comparatively better due to the availability of groundwater and accessibility to basic agricultural facilities.

3.3. District Ratlam: *An overview*

Ratlam is situated in the westernmost portion of the state of Madhya Pradesh surrounded by the districts of Dhar, Jhabua, Ujjain and Mandsaur. The district extends from 23^0 05' N to 23^0 52' N and 74^0 31' E to 75^0 41' E. The total Area of Ratlam District is 4861 Sq.km. which is 1.11% of the total area of Madhya Pradesh. The average rainfall of the district is 90 cm. Most of the rains occur in the month of July and August. The average number of rainy days is 42.3 over a period of 40 years. The forest cover of Ratlam is 1.19 % of the geographical area. The extent of wasteland in the district is considerably high and covers 32.50 % of the total geographical area of the district. These wastelands mainly include gullied uplands and ravines, undulating slopes with or without scrub, degraded forest land, degraded land under plantation crop, inland sands, and barren rocky land and sheet rock area. The district also has high poverty and a considerable Scheduled Tribe (ST) population (25.89%). The major economic activities in Ratlam include small and medium scale industries like chemical factories, copper wire and plastic rope manufacturing, and agricultural activities. The agriculture is mostly rainfed and the major crop of the district is soya bean, wheat, gram and maize. Besides, millets like jowar, other pulses like udad, oil seeds like groundnut, and cash crops like cotton are also grown. The average groundwater depth recorded in various sampling points across the district by the Central Ground Water Board over a period of five years is 38 feet. This district is also characterized by similar factors as the Dhar district in terms of high poverty, considerable ST population, low rainfall, critical groundwater, undulating terrain and low forest cover, all of which make this district suitable for the current study.

3.3.1. Selection of Blocks

Jaora and Bajna blocks of the Ratlam district were selected for study. The total population of Jaora and Bajna block is 243070 (male-123843 and female-119227) and 80290 (male-40361 and female-39929) respectively. The scheduled Caste (SC)/ST population of Jaora is 21.78/03.97% and

Bajna is 0.70/95.22%. The literacy of the Bajna block (34.22%) is much lower than the average literacy of the district, i.e., 49.56%. The literacy of Jaora block (62.89%), however, is higher than the average literacy of the district. The prime profession of the villagers is agriculture and the average landholding in the block is 0.30 hectare. The agriculture of the Bajna block is constrained by a number of issues such as scarcity of irrigation, undulating topography, unfertile land, critical ground water level, inaccessibility to agricultural facility, and lesser awareness. On the other side, agriculture of the Jaora block is comparatively good due to better accessibility to agricultural facilities, fertile land and awareness.

3.4. Sampling Strategy

The selection of suitable sites is an important criterion to assess the vulnerability and their impacts on livelihood of rural communities. The main purpose of the study is to understand the feasibility of future adaptation initiatives and to formulate recommendations (for planners and policymakers) to improve individual and collective assets. The Malwa region was selected for the study because it is known for pockets of poverty and is inhabited by a higher concentration of scheduled tribes population (Table 3.1). The poverty is further aggravated by limited livelihood opportunities, lesser literacy, subsistence agriculture, degradation of natural resources, scarcity of water (surface and ground), deprivation of special government programmes and the absence of a strong people's organization /civil society/ NGO movement in the region. In the Malwa region, Dhar and Ratlam districts were selected for the study. These two districts have the highest tribal population and are known pockets of poverty.

Plate-1 a & b: Interaction with villagers during PRA.

Table-3.1: Percentage of ST population, Forest cover and Literacy in Districts of Malwa-Madhya Pradesh

S. No.	District	% ST Population	% Forest Cover	Literacy
1	Dewas	16.45%	27.04	70.50
2	**Dhar**	**54.50%**	**9.00**	**60.60**
3	Indore	6.65%	18.11	82.30
4	Ujjain	3.11%	0.49	73.60
5	Rajgarh	3.78%	2.49	62.70
6	**Ratlam**	**25.89%**	**1.19**	**68.00**
7	Mandsaur	0.031%	4.7	72.70
8	Shajapur	2.74%	0.47	70.20

In the Dhar district, Dhar and Gandhwani blocks were selected under non tribal and tribal categories respectively for the sampling. These blocks were selected on the basis of population of scheduled tribes and literacy. The Dhar block is in the non-tribal category because of lower ST population (36.44%) and higher literacy (57.18%), as compared to the other blocks of the district. On the other side the Gandhwani block is selected in the tribal category because of its higher scheduled tribal (90.74%) population and lower literacy rate (34.83%) [Table 3.2.] In Ratlam district the Jaora block (03.97%) is selected for sampling in the non- tribal category. On the other side the Bajna block, comprises 95.22% the tribal population and is selected under the tribal category. The criterion for selection is the same as the previous one.

Table-3.2: Data pertaining to selection of blocks in Dhar and Ratlam district

S.No.		Dhar		Ratlam	
		Dhar	Gandhwani	Jaora	Bajna
1	% ST Population	36.44	90.74	21.78	0.70
2	% SC	09.25	01.47	03.97	95.22
3	Literacy	57.18	34.83	62.89	34.22
4	% Working Population	43.00	48.18	46.03	51.14
5	Average Landholding	0.20	0.22	0.422	0.214

Source: Census of India (S.No. 01 to 04) and District Statistical Book (S.No.5)

Under each block ten villages were selected randomly. Villages with more than 90% tribal population in the tribal rich category, i.e., Gandhwani in Dhar and Bajna in Ratlam, and villages with less than 10% tribal population in the tribal poor category, i.e., Dhar block in the Dhar district and Jaora block in the Ratlam district, were selected for sampling (Table 3.3). Further, ten respondents were randomly selected from each village for personal interviews. Full enumeration of each village has not been possible due to similar responses toward climatic change and time constraints. The basic design undertaken to complete the study was collection of primary data from field surveys and secondary data from the concerned government authority. The intention of the data collection was to comprehend the villager's perception toward climate change, their impacts on livelihood, and coping (traditional and existing) mechanisms.

Table-3.3: Percentage of ST population in selected villages of Dhar and Ratlam District

	Dhar				Ratlam			
S. No.	**Dhar**	**% ST**	**Gandhwani**	**% ST**	**Jaora**	**% ST**	**Bajna**	**% ST**
1	Saktali	00	Jalokhiya	99.1	Rojana	7.3	Maanpur	100
2	Tornod	3.839	Machhlai	99.73	Banbara	20.55	Chotiya Bawdi	100
3	Samar	00	Malgarh	100	Bhuteda	5.51	Gadikatara Khurd	100
4	Musawada	0.135	Gungidevi	99.77	Uplai	0.64	Mandaliya	100
5	Antrai	4.115	Luharpura	100	Kerwasa	1.18	BadanKhurd	99.89
6	Lohari Bujurg	00	Kharcha	100	Dodiyana	12.49	KarbalKhora	99.79
7	Sirsoda	8.648	Panchpipilya	99.75	Udyakhedi	38.37	Lambisada	99.58
8	Baggad	0.156	Baledi	97.45	Patakhedi	7.19	Lalpur	99.87
9	Sukheda	1.251	Motimali	100	Rewas	0	Mahuripada	100
10	Nekpur	1.775	Kesavi	93.41	Badodiya	6.28	Jharaniya Ukala	100

Source: Census of India.

Plate-2 a, b, c and d : Personal interview and PRA.

A well-designed survey tools was used for evaluation and it helped in capturing both the qualitative and quantitative dimensions of data. These included semi-structured interviews, Participatory Rural Appraisal (PRA), Village Schedules, Focus Group Discussion (FGD), and Household Questionnaires. Separate FGDs were organized with various sections of the community: - landless and SC/ST, small and marginal farmers, women's groups and Panchayat members. For each key issue, a set of primary tools and secondary tools was selected from the above list. Primary tools were the main source of information and reflected the emphasis on data analysis and presentation. Secondary tools were seen as complementing/ substantiating/ triangulating information obtained through primary tools. Besides these, key informants such as elderly people, youth, people in government service, and village representatives were also interviewed. The various subsets of questionnaires were designed according to pre-selected indicators of different sectors for generation of information. During PRA, villagers were briefed about climate change, its impacts on livelihood, and a few coping strategies. The information and data regarding agricultural practices, forest cover, water availability, health, education and communication that contribute directly or indirectly to the overall development of villagers was also discussed very intensely.

4 Analytical Framework

The formulation of local vulnerability indicators as a benchmark is useful for transparent and efficient allocation of limited resources, despite the weakneses and flaws of these indicators. Local vulnerability assessment is based on the qualitative analysis of climate change impacts using a matrix of the participatory, as well as scheduled, interview process. The so-called "bottom-up" approach has been commonly used for local vulnerability assessment. The approach is of great use in formulating sector specific policies and programmes. The vulnerability of the Malwa region was assessed with the help of indexes which were composed of indicators and segregated into components. Figure 4.1 shows the explanation of all the components in the vulnerability assessment framework.

4.1. Identification and Selection of Indicators

Indicators constitute of information that is readily available, or can be obtained easily. An indicator has been defined as a variable and a measure of system behavior in terms of meaningful and perceptible attributes (Holling *et al.*, 1978). Further, the approach is used by a number of authors as: sign - defined as something which stands for something to somebody in some respect or capacity (Marcus, 1983); proxy - summarizes by a specific phenomenon (McQueen and Noak, 1988); a statistical measure (Tunstall, 1992); parameter - provides information on state of phenomenon and also defines it as value that is measured or observed (OECD, 1993); a piece of information (Bakkes *et al.*, 1994); index (Hammond *et al.*, 1995); variable - function from observable variable (Gallopin, 1997); sub-index or component of an index (Ott 1978; Hahn *et al.*, 2009; Pandey *et al.*, 2011).

A variable is an operational representation of an attribute (quality, characteristic, property) of a system and generally a variable 'indicates' an attribute (Gallopin, 1997). The purpose of the study is to identify the indicators, components and sectors of vulnerability. The indicators were chosen based on reviews of existing literature; preliminary field visits and consultation with experts. Indicators are case specific and may be qualitative characteristics with the potentiality of being expressed as quantitative factors through the use of proxies, as elaborated by Deems (2010). The selected indicators are transparent and easily observed by communities. A cluster or set of indicators is a component and it usually simplifies and summarizes indicators. The component comprises a range of interacting indicators of climate change. The components are measured in a systematic process of assigning a value to an indicator such as agricultural land, irrigation and yield and is represented in numerical form. A similar approach has also been followed by the United Nations Development Program (UNDP) in 2002 for its Human Development Index; Sullivan in 2002 for calculating the water poverty index; Hahn *et al.,* in 2009 for assessing livelihood vulnerability; Pandey *et al.,* in 2011 for measuring climate vulnerability; Esteves *et al.,* 2013 for assessing the potential of MGNREGA in vulnerability reduction and Sinha *et al.,* 2015 for evaluating the potential of incentive based plantation.

Step 1: Identification of components and scale (village level) – Components are region specific and identified through expert consultation and literature review.

Step 2: Identification and definition of indicators – Indicators are site specific and selected after preliminary field visit, expert consultation (Academicians, Researchers, District and block level staffs), and literature review. These were further refined based on autocorrelation considerations, expert judgment, baseline considerations, limitations and proxies.

Step 3: Quantification of indicators - Primary and secondary data are quantified. Qualitative as well as quantitative form of data is used and qualitative data is converted into quantitative form with the help of ranking.

Step 4: Normalization of indicators to a dimensionless unit – The numerical values of indicators are in different units (ha, meters, percentage, tonnes, etc.) and in order to compare them, they are normalized and rendered dimensionless for aggregation.

Step 5: Construction of indices – The numerical values of indicators are aggregated into indices.

Step 6: Aggregation of indicators to obtain vulnerability score – The indices are aggregated to obtain vulnerability score.

Step 7: Identification of Vulnerable sector – Indices with higher vulnerability score within each sectors.

Step 8: Capacity assessment - Key issues along-with existing initiatives (policies and programmes) are reviewed and priority sectors (sector with higher vulnerability) are identified. Further the approach for reducing vulnerability and enhancing adaptive capacity are also suggested.

Step 9: Prioritisation of components: Dimension reduction technique is used for prioritisation of components.

Fig. 1: Approach adopted for vulnerability and capacity Assessment.

4.2. Normalization of Indicators

The information on different indicators has been collected from households as per defined methodology (Table-4.1). A total of 80 indicators were identified for vulnerability assessment. The indicators were in different units and scales and thus some form of standardization was required to ensure that all the indicators are comparable. The indicators were standardized on the basis of functional relationship between the indicators and vulnerability so that the values are free from all units. Two types of functional relationship was possible: vulnerability increases with increase in the value of the indicator, i.e., positive relationship *(equation-i);* and vulnerability decrease with increase in value of the indicator, i.e., negative relationship *(equation-ii).* It is interpreted that higher the value of the indicator, more is the vulnerability.

$$\text{Index}_{sv} = \frac{S_v - S_{min}}{S_{max} - S_{min}} \quad \text{.................................}(i)$$

$$\text{Index}_{sv} = \frac{1}{(S_v + 1)} \quad \text{.......................................}(ii)$$

Where,

S_v is the average value of the indicator at the respondent level.

S_{min} and S_{max} are the minimum and maximum values of the indicator at the respondent level.

After normalization of each indicator they were averaged using equation-(iii) to calculate the value of each index.

$$M_v = \frac{\sum_{i=1}^{n} Index}{n} \quad \text{......................................}(iii)$$

Where,

M_v is the indexes one of the component,

Index is the indicator value of ith indicator

n is the number of indicators in the index.

Each indicator contributes equally to the overall index, as also considered in other indices such as Hahn *et al.,* 2009 and Pandey *et al.,* 2011. These components have been directly and indirectly, and linearly and non-linearly related with the individual family's inherent quality, resources availability and accessibility. The indicators' values have been further aggregated into the components and converted into dimensions of vulnerability. These were Adaptive capacity, Sensitivity and Exposure. The CoV, further tested statistically (T-test, ANOVA and PCA) to analyze the difference between tribal and non tribal blocks and factors with the highest variance.

Table-4.1: Indices of Exposure, Sensitivity and Adaptive Capacity

Components	Exposure	Sensitivity	Adaptive Capacity
Socio-Economic Profile		• Family dependency index • Poverty line index • Migration index	• Profession diversity index • House type diversity index • Asset index • Livestock index • Electricity index • Sanitation index • Stability index
Social Network			• Assistance index • Money transaction index
Education			• Youth education index • Household education index • Household decision index
Agriculture		• Cropping intensity index	• Agricultural equipments index • Irrigation index • Agricultural land index • Irrigation strategy index • Agricultural strategy index
Food			• Food sufficiency index • Nutritious food index
Water		• Potable water availability index	• Potable water Accessibility Index • Potable water collection index
Weather Perception and Calamities	• Summer temperature perception index • Rainfall perception index • Winter temperature perception index • Natural Calamities index • Stress Index • Climatic disease index		• Weather forecast index
Natural Resources	• Natural resource degradation index	• Natural resource availability index • Natural resource collection index • Natural water quality index	• Fodder strategy index • Fuelwood strategy index

Health	• Health worker index • Health facility accessibility index
Infrastructure	• Connectivity index • Market accessibility index • Bank accessibility index • Government resource index

4.4. Identification of Components

The component of vulnerability (CoV) is a set of indices that measure vulnerability by aggregating indicators. It comprises the crucial sectors of well being, that are directly or indirectly impacted by climate change, namely, socio-economic profile; social network; education; health; infrastructure; natural resources; agriculture; food; water; and weather perception and calamities. Each CoV contains relevant indicators based on specificity of the region and selected by the interactions with the peers, locals of the region, block and district level officials, and a review of the literature.

The majority of the world's poor live in rural areas (Global Monitoring Report, 2013), and their socio-economic status is not good due to widespread poverty and inaccessibility to basic life supporting facilities. Therefore the socio-economic status of a region is a central indicator for vulnerability and capacity assessment. The socio-economic profile includes information pertaining to number of family members; profession of household; house type and assets; livestock; migration; electricity; sanitation, and stability.

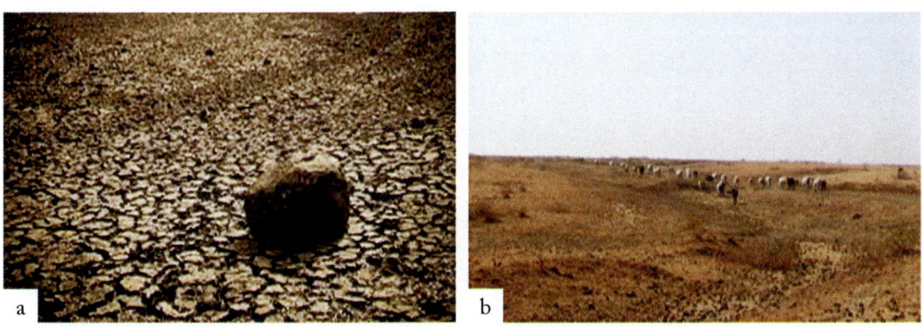

Plate-3 a & b: Impact of Climate change on Rural Agriculture.

Strong social network is the potential of a region for social support such as money, food, and non-food items. Therefore, evaluation of social support is important for formulation of effective adaptation programmes. The indicators of social network are context specific, and range from willingness to wealth of villagers in providing/seeking assistance in monetary as well as non-monetary form.

Relevant, locally-controlled educational programs play key roles in rural development. Education imparts social change and has influence over poverty, income equity and employment. In the present study, educational status of the households was captured for analyzing the educational status of villagers; education among children and school drop-outs. In the long run, education influences the decision making ability of the households and hence the educational background of the head of household was captured and analysed.

Climate change is projected to have mostly negative and large health impacts on the rural poor. The health-care facility in rural areas of Malwa region is however, pitiable due to lack of quality infrastructure, shortage of medical functionaries and inaccessibility to basic medicines and medical facilities. Thus a proper assessment of health care at the village level is requisite for strengthening rural capacity against climate change. Indicators like frequency of visits of doctor and health worker and the distance of a village from the hospital and medicine shop are selected for analysis.

Rural infrastructure is not only a key component of rural development, but also an important ingredient in ensuring any sustainable poverty reduction programe. The proper development of infrastructure in rural areas improves rural economy and quality of life. It promotes better productivity, increased agricultural incomes, adequate employment, etc. But in a few tribal villages of Malwa region, infrastructural development is completely lacking. Thus there is an urgent need for facilitating rural areas by creating basic amenities. For understanding the different aspects of rural development, basic infrastructure at village level was analyzed through the availability of a panchayat bhawan; primary school; distance from bank and rural connectivity.

Forests provide a wide range of socioeconomic and social benefits to the rural poor. A section of rural population depends heavily on natural resources that are under tremendous population pressure and also affected by climate change. The present study examines the potential of forests in satisfying the requirements of fodder and fuel-wood in villages. In the Malwa region, a large section of population primarily depends on forests for their livelihood.

Plate-4 a: Maintenance of roads connected to tribal villages is very less.

Plate-4 b: A few tribal villages still not connected with all weather road.

The unsustainable harvest causes destruction and degradation of forests that ultimately affect the rural poor in the long run. Therefore the present status of forests was assessed in tribal and non tribal villages with the help of indicators like the time spent for collection of forest produce and change in distance between forest and village.

Plate-5: Fuelwood collection from forest.

Agriculture is considered susceptible to climate change because of the impact of temperature and precipitation. It is however, the prime source of income and livelihood in rural areas. Climate change related impacts will progressively increase over this century and it is important to analyze the current agricultural vulnerability of the Malwa region so that an appropriate programme/policy would be formulated. The agricultural vulnerability of the region is captured with the help of indicators like total agricultural land; crop yield; cropping intensity; irrigation facility, and strategies of households for improving agriculture.

Climate change would adversely impact food production, availability and accessibility. Those households that are already unable to maintain food

Plate-6 a: Agriculture of non tribal villages. **Plate-6 b:** Agriculture of tribal villages.

security are the ones that are most sensitive to climate variability (Maxwell and Smith 1992). To assess the food security of households, various indicators were used, a central one being the primary source of food, food sufficiency to feed all household members throughout the year, and variation in consumption of nutritious food in the last ten years. In both blocks, agriculture is the prime source of food, although a few of them also depend on Public Distribution Systems (PDS).

The accessibility to adequate and safe drinking water is vital for human health, well-being, economic production and sustainable development. Inaccessibility to drinking water may expose the community to the risk of outbreaks of waterborne and infectious diseases. Further, the additional stress from climate change impact will require significant adaptation action. Thus the accessibility to safe drinking water in the Malwa region was taken as a component for vulnerability and capacity assessment. The assessment was carried out with indicators like source of drinking water in the village; number of months of availability of water in the sources; and time spent and distance travelled for water collection.

Plate-7: Drinking water collection in lean/summer season in tribal villages is difficult.

Rural communities are particularly more exposed to climate change and variability because of specific geographic and demographic barriers such as physical isolation, limited livelihood opportunity, higher poverty and population pressure on limited resources, even though rural people are good observers of variation in climate phenomenon. To get more about the perception of rural communities on climate change and their traditional practices, weather perception and calamities are chosen as components. Data on intensity and frequency of extreme weather events and stress due to climatic variability were collected to analyze the impact of extreme events.

4.5. Vulnerability Assessment: IPCC Approach

The Intergovernmental Panel on Climate Change's vulnerability framework breaks down the concept of vulnerability into dimensions of exposure (external disturbances); sensitivity (change brought about by exposure) and adaptive capacity (ability to adjust itself to minimize the damage). The greater the exposure or sensitivity, the greater is the vulnerability. Adaptive capacity is however, inversely related to vulnerability. So, the greater the adaptive capacity, the lesser is the vulnerability. The Vulnerability and Capacity Assessment (VCA) index is calculated on the basis of the IPCC's vulnerability assessment framework. Wherein, exposure and sensitivity together describe the potential impact that climate change can have on a system and the adaptive capacity is generally accepted as a desirable property or positive attribute of a system for reducing vulnerability (Engle, 2011). In the VCA framework, vulnerability is calculated by estimating dimensions of vulnerability (DoV), i.e., adaptive capacity; and sensitivity and exposure through an indicator based approach.

4.5.1. Adaptive Capacity

Adaptive capacity is a significant factor in characterizing vulnerability. In the context of climate change, it is closely related to concepts such as adaptability, coping ability, management capacity, stability, robustness, flexibility, and resilience (Smit and Wandel, 2006). According to Brooks (2003), the adaptive capacity of a system or society reflects its ability to modify its characteristics or behavior in order to better cope with existing or anticipated external stresses and changes in external conditions. The IPCC (2014) in its fifth assessment report, describes adaptive capacity as the potential or ability of a system, region, or community to adjust to actual or expected climate and its effects. The natural systems have the potential to adapt through multiple autonomous processes (e.g., phenology changes, migration, compositional changes, phenotypic acclimation, and/or genetic changes), and humans may intervene to promote particular adjustments such as reducing non-climate stresses or through managed migration. But successful adaptation will depend on our ability to allow and facilitate natural systems to adjust to a changing climate, thus maintaining the ecosystem services on which all life depends. The capacity to adapt is context-specific and varies from country to country, from community to community, among social groups and individuals, and over time (IPCC 2001; Smit and Wandel, 2006). Adaptive capacity is considered to be "a function of wealth, technology, education, information, skills, infrastructure, access to resources, and stability and management capabilities" (IPCC, 2001; McCarthy *et al.*, 2001). There are close linkages between adaptive capacity and livelihoods or asset ownership. The more assets people have, the higher adaptability they have; conversely, the lesser the erosion of people's assets, the greater their security. In the study, adaptive capacity illustrates the internal as well as external capacity, or capability, that reduces the impact of extreme events. We scrutinized ten sectors to fabricate existing adaptive capacity with the help of indicators such as profession diversity index; house type index; livelihood index; sanitation index, etc.

4.5.2. Exposure

Exposure refers to the inventory of elements in an area in which hazard events may occur (Cardona, 1990; UNISDR, 2009). Hence, if population and economic resources were not located in (exposed to) potentially dangerous settings, no problem of disaster risk would exist. Typical exposure includes variation in temperature and precipitation. Changes in these parameters can exert major additional stress on natural as well as anthropogenic systems. In the context of the present study, exposure is a potential impact of climate change. The impact of climate change is not measured at the village level. Therefore, the perception of respondents for different parameters of climate is considered as exposure.

4.5.3. Sensitivity

Climate sensitivity describes how sensitive the global climate is to a change in the amount of solar energy reaching Earth's surface and lower atmosphere.

For example, the amount of carbon dioxide in Earth's atmosphere has doubled from the pre-industrial level (280 ppm to 560 ppm), this will cause an energy imbalance by trapping more outgoing thermal radiation in the atmosphere. Sensitivity is the degree to which a system is exposed by an internal or external set of disturbances. This measure, which reflects the responsiveness of a system to climatic influences, is typically shaped by natural and/or physical attributes of a system (topography, erosion, etc.). But it is also altered or changed by anthropogenic activities which affect the physical constitution such as tillage systems, water management, resource depletion and population pressure (GIZ, 2014). In the present study framework, sensitivity describes the human–environmental conditions that can either worsen the hazard or trigger an impact. Sensitivity in the Malwa region is measured by assessing the current state of poverty; family dependency; cropping intensity and irrigation, etc. (Table. 4.1).

4.6. Vulnerability and Capacity Assessment (VCA)

VCA is an indicator based bottom up tool for vulnerability and capacity assessment. It captures the dimensions of livelihood that are directly or indirectly associated with the impact of climate change. It is mainly conducted at the grass root level, focusing on the experiences and perceptions of stakeholders regarding climate change. It intrinsically captures the data and information on livelihoods of people and is a potentially powerful tool for designing capacity enhancement measures at the local level and in supporting the creation of wider mitigation measures. The purpose of VCA is to address the impacts of climate change on the livelihoods of rural communities, their perception of these changes, and their capacity to adapt. It encourages involvement of affected people through community based investigation processes like Participatory Research Approaches (PRA) and personal interviews.

VCA is constructed to improve understanding of how environmental and socioeconomic changes affect the rural livelihoods, what shapes their vulnerabilities, and what assets they have for coping with and adapting to environmental and socioeconomic changes. In totality, VCA is about collecting, analyzing and systematizing information on a given community's vulnerability in a structured and meaningful way. The captured information is used to diagnose the key risks and existing capacities of the community, ultimately leading to activities aimed at reducing people's vulnerability to potential disasters and increasing their capacity to survive them and resume their lives. It enables local priorities to be identified and leads to the design of actions that contribute to disaster reduction, as well as the design and development of programmes/policies. This approach also brings some inevitable gaps and limitations that should be designed before the planning process. This helps in evading the downsides associated with secondary data and is critical in reducing the dependence on climate models (Hahn, 2009) for planning adaptive capacity in rural communities. VCA is an integral part of community based capacity enhancement programs and plays central role in reducing the impact of extreme events where possible (through mitigation, prediction/warning and preparedness) and enhances capacity by diagnosing communities' root causes of vulnerability. It is an accumulation of differential dimensions of vulnerability, which is a prerequisite for the formulation and execution of equitable and sustainable programms/policies. Further, it can accurately assess the relationships across a wide range of stressors because it uses primary data from household surveys at the village level for constructing the index.

5 Results and Discussion

Over the last few decades, Madhya Pradesh has witnessed a radical transformation in terms of economic and social development. The state has recorded an impressive growth in the agricultural sector (24.99 % in 2013-14) and infrastructural development, supplemented by the creation of robust institutions. The CSO has estimated an impressive 10.19 % Gross State Domestic Product (GSDP) growth rate of Madhya Pradesh in the year 2014-15. As per the quick estimates, the GSDP of Madhya Pradesh grew by 9.48 % during the year 2013-14 against the projected 11.08%. The actual growth rate in 2012-13 was 9.9%. Further, the abundant manpower of the state has been developed into a skilled technical workforce resulting in the per capita income of the state increasing significantly (Rs 59,770 -2014-15 from Rs 15,442 -2004-05) (www.freepressjournal;www.dailypioneer.com). In contrast, the state's performance on basic human development indicators (0.394) is not impressive, and many a time fall far below the set standards or the national average Maternal and Paternal Death Review (MPDR), 2011. Key sectors such as health, access to safe drinking water and sanitation, housing and electricity are still causes of great concerns if the state has to pursue a balanced and equitable socioeconomic development. Thus, the state has to put forth sector specific strategic efforts in improving the current pace of providing services. In the light of the above, a research study on VCA was conducted in the Malwa region of Madhya Pradesh to identify the dimensions of development for enhancing capacity of rural communities, and thereby, the emerging policy thrust for development. The result is presented in different sections. Section I presents the components of vulnerability, their impacts on livelihood of rural communities and categorization of the vulnerable social groups. The perception of rural communities toward climate change and identification of community-based coping mechanisms is also assessed. Section II analyzes village wise vulnerability with the help of the IPCC vulnerability assessment framework. The concluding recommendation is designed with dimension reduction techniques and is discussed in section III.

5.1. Components of Vulnerability and their Impacts on Livelihood

5.1.1. Socio-Economic Profile

The socio-economic profile takes into account the household's composition as well as its socio-economic status. A central indicator of socio-economic status is the educational background of the household, which is measured by the level of education of the household head. The head of the household is of prime importance in regard to resource management, strategic planning, and decision making at the household level. Apart from education, other indicators like poverty, migration, sanitation and access to electricity also play a critical role in socio-economic analysis. Test of significance using a t-test was done to make sure that there is a statistically significant difference between tribal and non-tribal communities (Annex 8). The socio-economic status of the region was measured with the help of ten indices composed of 12 indicators. Most of the indices of tribal and non tribal blocks were significantly different. Only the livelihood index does not show a significant difference because of almost similar livelihood vulnerability in all the sampled blocks. Further, the socio economic status of tribal and non tribal blocks also showed significant difference. The variability among the tribes of two districts in the same components is also measured with help of Analysis of Variance (ANOVA) (Annex 9).

Plate-8 a & b: Traditional tribal houses.

In Dhar district, the socio-economic profile of Gandhwani (0.597) block, which represents the tribal block in the study, shows higher vulnerability as compared to Dhar (0.267) block (Annex 2 & 3), the non-tribal block in the study. The same trend was also observed in Ratlam district, where the tribal (Bajna - 0.594) block showed higher socio-economic vulnerability as compared to the non tribal (Jaora - 0.209) block (Annex 4 & 5). The higher vulnerability of tribal blocks might be attributed to high family dependency, lesser opportunity to work in different professions, high poverty, high proportion of households having non-concrete houses, less household assets, indigenous livestock, high migration rate, low proportion of villages having electricity availability of more than 10 hours/day, and poor sanitation facility.

The dependency index of a household is calculated as a percentage of household members less than 18 years and above 55 years. In general, members belonging to this group don't contribute to the household income and, therefore, they are considered as dependent. Along with this, the group is susceptible to climate change impacts through direct and indirect ways. The dependency index of Gandhwani block is 0.486 and Dhar block is 0.398. On the other side, the dependency index of Bajna and Joara blocks of Ratlam district is 0.512 and 0.410 respectively. The higher vulnerability score of Gandhwani block of Dhar district and Bajna block of Ratlam district might be attributed to large family size and more number of dependents per household. The livelihood diversification for income sources have been the most widespread adaptations of rural people of the Malwa region. Rural livelihoods diversification has generally occurred as a result of an increased importance of off-farm wage labor in the household livelihood portfolio or through the development of new forms of on-farm/on-site production of non-conventional marketable commodities. The livelihood diversification is higher in non-tribal blocks (Dhar and Bajna) due to more opportunity to work in different professions. The prime profession of the villagers is agriculture and wage labour. At least two to three members of each household are involved in agriculture as farmers or wage labourers. Additionally, a few were involved in government/private jobs; business (grocery/chicken/mutton shop) and own skill job. The diverse and strategic livelihood attempts to optimize household capacity to take advantage of ever-changing opportunities and cope with unexpected constraints. On the other side, the livelihood diversification is lower in tribal regions

(Gandhwani and Bajna). In tribal regions there is a possibility of enhancing livelihood strategies by keeping the capability to operate a heterogeneous set of activities. The diversified livelihoods of households may reduce the vulnerability and enhance the adaptive capacity of the respondents.

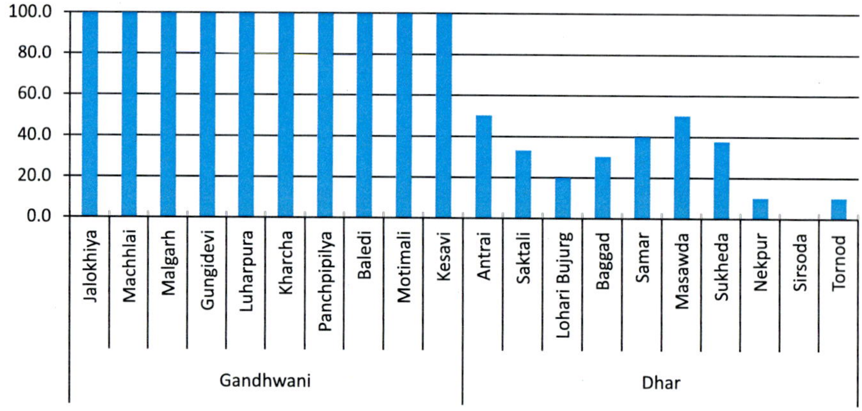

Fig. 5.1.A: Percentage of BPL households in the sampled villages of Dhar and Gandhwani blocks of Malwa Region.

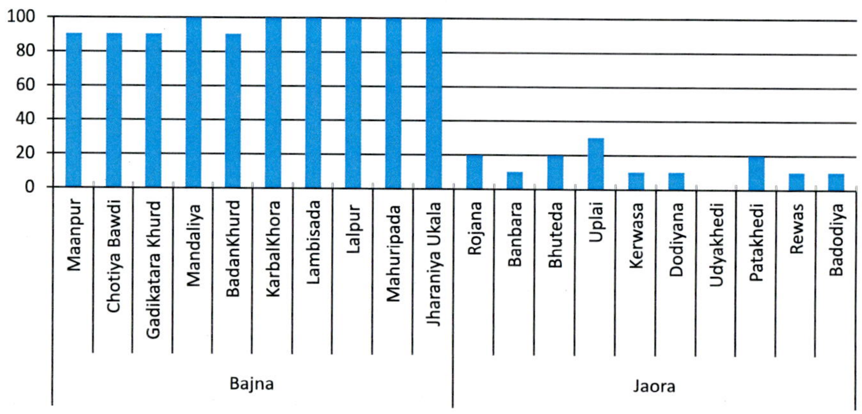

Fig. 5.1.B: Percentage of BPL households in the sampled villages of Bajna and Jaora blocks of Malwa Region.

Source: Primary survey.

A house is the central indicator of the wellbeing of a household and is strongly linked to the human right to adequate housing. Housing fulfils physical, psychological, and social needs. It provides security and shelter from the weather and climate, a sense of personal space and privacy, and space for communal gathering (Human Rights Education Associates, 2012). One of the focuses during the interview was on collecting information about the types of houses. The types of houses were categorized into different classes, i.e., concrete, semi-concrete and non-concrete. In Madhya Pradesh, a greater section of the population having non-concrete houses, i.e., 61.30 % out of the total 71 % non-concrete houses are in rural areas. On the other side, only 18.50% households have concrete houses and the remaining 20.20% possess semi-concrete houses (2007-08) (IIPS, 2010). In the Malwa region especially in tribal areas poverty is very high; a large proportion of households belong to the BPL category and a majority of them possess non-concrete houses (Fig. 5.2.A&B). These houses are inapt in terms of sheltering them form extreme weather events. Further, government intervention like the Indira Awas Yojana and the Mukhyamantri Awas Yojana provide housing facility to the BPL families, but due to limited allocation as prescribed in the scheme, only a minority of the population is benefited by this. Therefore, vulnerability in tribal regions (Bajna - 0.446 and Gandhwani - 0.467) is higher as compared to non-tribal regions. The houses in non tribal region, i.e., Dhar (0.318) and Jaora (0.273) block are semi concrete and concrete; this can be attributed to higher household incomes and would be further justified by a higher proportion of Above Poverty Line (APL) population (Fig.5.1.A & B). The semi concrete and concrete houses provide better protection during extreme events that lead to lower vulnerability of the households. The difference in housing pattern in tribal and non tribal blocks of Dhar, as well as Ratlam district, is further tested statistically and the result shows that the index is significantly different in tribal and non-tribal blocks.

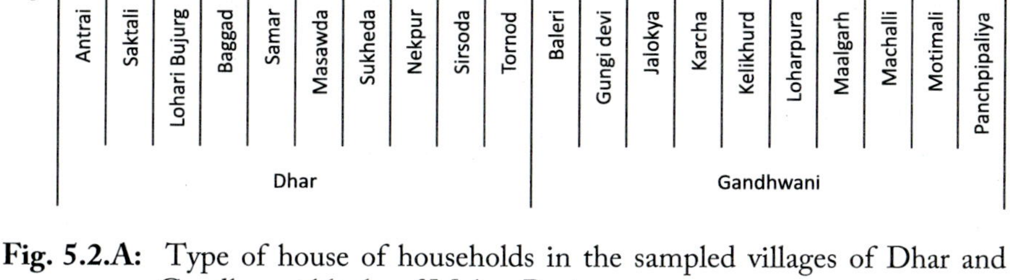

Fig. 5.2.A: Type of house of households in the sampled villages of Dhar and Gandhwani blocks of Malwa Region.

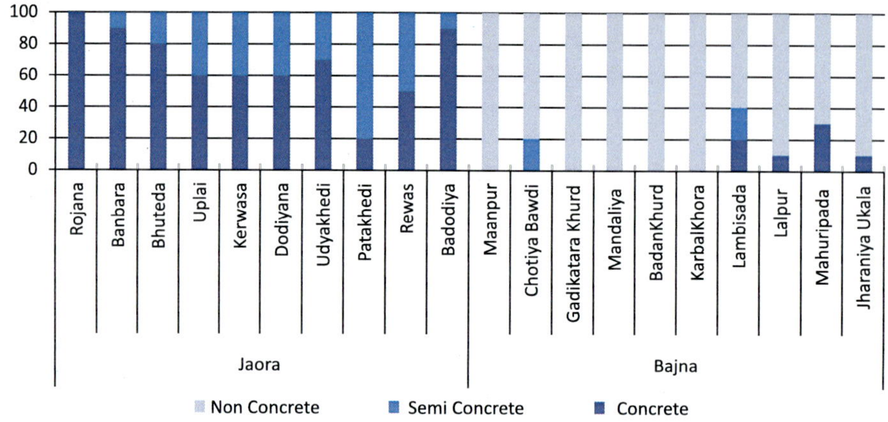

Fig. 5.2.B: Type of house of households in the sampled villages of Jaora and Bajna blocks of Malwa Region.

Source: Primary survey.

The possession of durable goods and assets is a central dimension of material wellbeing (Haughton and Khandker, 2009). It is a useful proxy for the socio-economic status of a household, especially if the secondary data on income or household consumption is not available or is unreliable (McKenzie, 2005). In the questionnaire, this dimension is measured by the household's ownership on assets. Households of Dhar block possess more durable assets like cycles, motorcycles, tractors, mobiles, radios, television sets and motor pumps. In non tribal block cycles, motorcycles, television sets and mobiles are very common and many of them also possess tractors and motor pumps. Therefore, the assets vulnerability of Dhar (0.222) and Jaora (0.153) is lower compared to tribal blocks (Gandhwani – 0.623 and Bajna – 0.336). The higher proportion of population of Gandhwani and Bajna block possesses only cycle and mobile and a few of them have motor cycles, television sets, tractors and motor pumps. The higher asset vulnerability index can be attributed to lesser income generating opportunity, distance to market and daily supply of electricity.

Livestock integration with livelihood might be a part of strategies to assist the construction of sustainable livelihoods, especially in rural settings. It is, therefore, essential to be aware how different livelihood strategies complement one another. In the Malwa region, livestock integration in livelihood incomes especially among tribal blocks, enhances adaptive capacity. Livestock not only provides nutritious food such as milk but it is also a source of income for them. The number of livestock per household in tribal blocks was high. They rear livestock as a traditional practice. Approximately each household had two or three livestock but they possessed indigenous breeds. This leads to low milk production and higher pressure on forest for fodder. On other side, in non tribal blocks, all the households do not have livestock, but most of them possess hybrid livestock. The livestock vulnerability was calculated based on the number of livestock per household. Therefore, non tribal blocks showed slightly higher vulnerability. During interactions, households of tribal blocks showed their inclination for rearing hybrid livestock and they added that due to lack of facilities (quality fodder, dwelling for the animals, veterinary doctor and market for selling of products) and high investment, they do not prefer hybrid livestock.

Plate-9 a : Livestock rearing in Malwa region.

Plate-9 b : Livestock rearing in Tribal villages.

<u>Box 5.1: Case Study on</u>

<u>"Poverty and extreme weather events hit the poor the hardest"</u>

Mr. Jhuria of Karcha village of Gandhwani lives with his wife and six children in a non-concrete dwelling. He owns about five *bigha* of agricultural land and two livestock. Agricultural production is less and for sustaining his livelihood he works as a daily wage labourer.

In January 2011, he had taken a loan of Rs. 25,000 from a local lender to purchase a hybrid buffalo. He thought selling milk may contribute to his monthly income and help in reducing overall poverty. Unfortunately, due to heavy and incessant rainfall, along with improper dwelling facility, the buffalo died right after a chronic illness.

The monthly income of Jhuria from all sources is about Rs. 2500 to 3000 and unfortunately he is not able to repay the loan amount. Presently, he is bound to pay an interest amount of Rs. 250 per month.

In the above situation, he is unable to provide two meals per day to his family. The children are wholly dependent on the *mid-day meal* provided by the school. Sometimes, he has to sustain himself on one meal a day.

Poverty, coupled with the unavailability of work, was the main reason for migration in most of the villages. For example, more than 80% of the respondents of tribal blocks stated that lack of work is a cause of migration. In some cases, the shortage of food and repayment of debt were also the factors responsible for migration. The dominance of a large number of marginal farmers and landless labourers, coupled with a poor resource base, poor agrarian economy and agricultural failure, compel them to migrate in search of livelihood. Sometimes, the whole family has to migrate in order to earn a living. Migration is mostly observed for unskilled labourers. On the other side, migration of skilled labour is very less; possibly they get employment in villages or in the local market. Most of the migrant households migrate to rural areas, where work is available, though some of them migrate to urban areas for construction work or other informal sector work. Further, the mass migration is observed in some villages of Gandhwani and Bajna blocks and it more or less depends on the success or failure of agriculture. When the agriculture was affected by drought or heavy rainfall, most of the households used to migrate in the search of livelihood.

Most of the cases of migration are found in the tribal block. The migration in this block is temporary; usually people migrate for six to eight months and return in the subsequent agricultural season when they observe optimum conditions for agriculture. The higher migration results into higher vulnerability of Gandhwani (0.654) and Bajna (0.950) blocks. On the other side, the lower vulnerability of Dhar (0.040) and Jaora (0.030) blocks can be attributed to low migration coupled with a better agricultural base, good work opportunity and comparatively lesser poverty (Annex 2,3,4 & 5).

Further, access to electricity has implications for the health, education, and income of households and communities. Among other things, electricity enables children to study at night and help to use information and communication technologies. The most important is irrigation that totally depends on electricity in the rural setting. In the questionnaire, electricity facility is measured by accessibility to electricity and availability (hour/day) of electricity. During the survey, it was found that, except for a few, all villages are electrified. The state also reported impressive progress for access to electricity. The percentage of electrified villages has been increasing significantly since 1991 (43.30% - 1991; 70% - 2001 and 97% - 2011). However, the percentage of households using electricity as their primary source of lighting decreased in the last decade (70.0% - 2001 and 67.1%

- 2011). This might be attributed to increased demand or higher rate of electrification and constant supply. The availability of electricity in Dhar and Jaora blocks is 12 to 18 hours/day while it is six to ten hours/day in Gandhwani and Bajna blocks leading to higher vulnerability.

Sanitation influences the incidence of infectious diseases (Tol *et al.,* 2007). Thus, access to sanitation is particularly crucial to build up preparedness to various natural disasters exacerbated by climate change (McMichael & Woodruff, 2005; Keim, 2008). The indicator is quantified by the proportion of the population with access to toilet facilities that can effectively prevent human, animal, and insect contact with excreta. In tribal blocks most of the households lacked toilet facilities. This led to vulnerability of the Gandhwani (0.93) and Bajna (1.0) blocks which is comparatively very high. A similar trend is also observed in the state, with only 28.80% (Census, 2011) households having toilet facility. The proportion of households having toilet facility further decreased to 13.2% in rural areas.

In tribal regions households do not possess insurance and hence the insurance vulnerability in Gandhwani and Bajna blocks is very high. On the other side, a large section of households in Dhar, and Jaora blocks possess life as well as crop insurance. This makes them more secured for life. Further, the secured families are more involved in their communities as home owners, voters and can take better decisions.

Key Issues

1. Migration in tribal areas is very high and temporary; usually, people migrate for six to eight months and return in the subsequent agricultural season when they observe optimum conditions for agriculture. Migration is prominent in tribal regions and in many villages the whole family, along with children, has to migrate in order to earn their living.
2. Accessibility to electricity is good in all the sampled villages, but availability of electricity (hours/day) especially in the agricultural season is less in tribal villages of Dhar and Ratlam districts.
3. Most of the households of tribal regions possess non concrete dwellings and indigenous livestock.
4. Majority of household in non-tribal villages possess toilet, however, they were especially males, not using toilet.
5. In tribal regions households had lesser opportunity to work in different professions and, hence had higher poverty.
6. Lack of awareness about life as well as crop insurance in tribal villages.

Existing/Ongoing Initiatives

1. Sanitation A) Swachh Bharat Abhiyan: The prime objective of the mission is to create sanitation facilities for all. It aims to provide every rural family with a toilet by 2019. B) Samagra Swachta Abhiyan: Sanitation and hygiene in schools.
2. MGNREGA aims to enhance livelihood opportunities in rural areas by providing at least a hundred day employment guarantee to an adult member of a rural household who is willing to volunteer to do unskilled manual work in a financial year. A large section of population in tribal, as well as non tribals region benefited from the program.
3. Provision of concrete dwellings for SC/ST/BPL households under Indra Awas Yojana and Mukhya Mantri Awas Yojana. Under the scheme, only a few households were benefited because of lesser allocation of budget.
4. Provision of sustainable livelihoods through the Integrated Water Management Programme (IWMP). This enables multi-cropping and the introduction of diverse agro-based activities, which help to provide sustainable livelihoods to the people residing in the watershed area. In addition, there is a Scheme of Technology Development, Extension and Training (TDET), also being implemented to promote development of cost effective and proven technologies to support watershed management.

5.1.2. Social Network

Social networks reflect the potential for social support. Obtaining loans (money, food, and non-food items) and assistance from formal and informal institutions during extreme events is an important strategy used by households to cope with environmental and socio-economic shocks. In the questionnaire, social capital is measured by indicators like assistance received/obtained during an extreme event, money transaction, source loan, and purpose of loan. Social network consists of two indices, which are composed of five indicators. All the indices of social networks were tested statistically (T-test and ANOVA) and they showed significant differences. In totality, social networks showed significant differences in tribal and non tribal blocks. Further, the variance in social networks in the tribal blocks of Dhar and Ratlam districts and the non-tribal blocks of Dhar and Ratlam was assessed with the help of ANOVA.

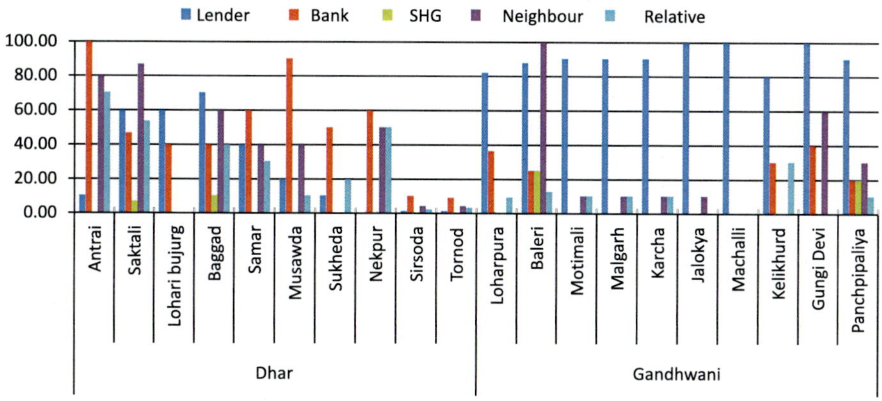

Fig. 5.3.A: Source of loan for the households of sampled villages of Dhar and Gandhwani blocks of Malwa Region.

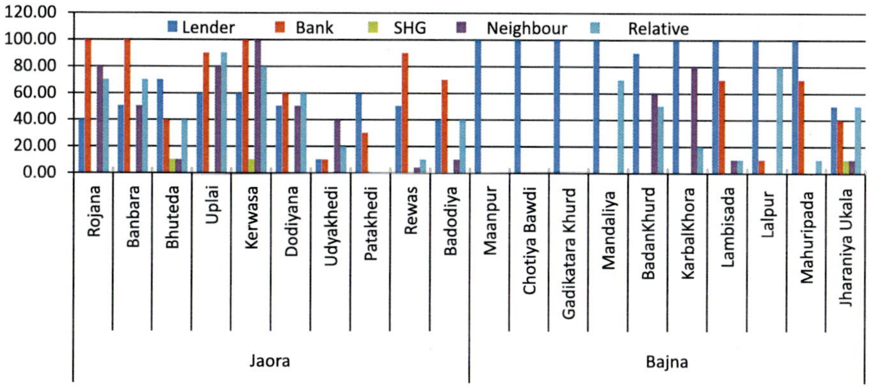

Fig. 5.3.B: Source of loan for the households of sampled villages of Jaora and Bajna blocks of Malwa Region.

Source: Primary survey.

Except for a few villages, households of Gandhwani and Bajna blocks do not acquire assistance due to poor access to formal and non-formal institutions. This makes households of these blocks lesser self-reliant and results in lesser capacity to provide assistance to others during extreme events. On the other side, households of Dhar and Jaora blocks have better access to formal and non-formal institutions and almost all the villages get benefited from these institutions. The availability and access to adequate, timely and low cost credit from institutional sources is of great importance, especially to small and marginal farmers. Most of the farmers in the study are small and marginal farmers engaged in agricultural activities of widely varying potential.

Money transactions amongst households is a very common practice in non tribal blocks. Households cooperate with each other in any emergency or extreme events. They help/support/co-operate with others in terms of financial and physical means. Apart from relatives and neighbors, they also get financial assistance from banks (Fig. 5.3.A&B). On the other side, money transaction is lacking in tribal blocks and inaccessibility to banks coupled with fewer co-operations among villagers compelled them to take loan from local lenders at a higher interest rate.

Most of the households in tribal and non tribal blocks are involved in agriculture, so the prime purpose of taking loans is agriculture. Apart from this, a few of them also take loans for medical and marriage purposes. A marriage ceremony in a tribal block is one of the important life events and hence they spent a large sum of money. Mostly they take loans and even mortgage their land for the marriage ceremony (Fig. 5.4.A&B). Further, the frequency of taking loans for medical purposes is again higher in tribal blocks due to their inaccessibility to better health facilities. Chronic poverty compels them to take loans for sustaining their livelihood. During discussions some of respondents added that villagers are forced to take loans from local lenders. If they do not take loans for agricultural purposes lenders refused to give loans during emergencies.

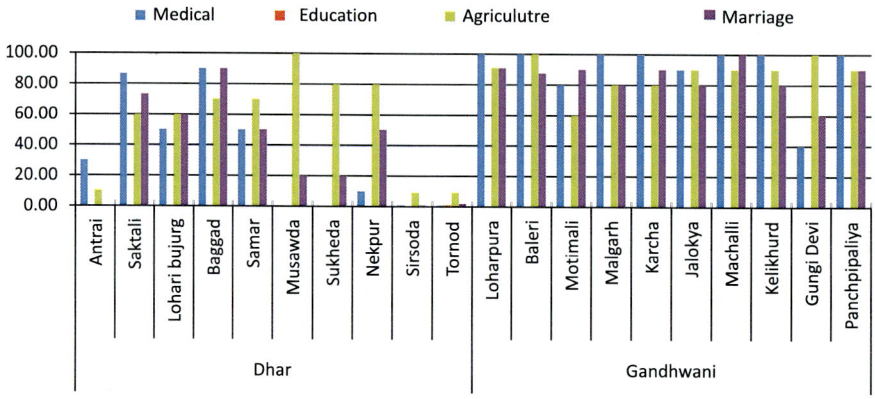

Fig. 5.4.A: Purpose of loan for the households of Dhar and Gandhwani blocks of Malwa Region.

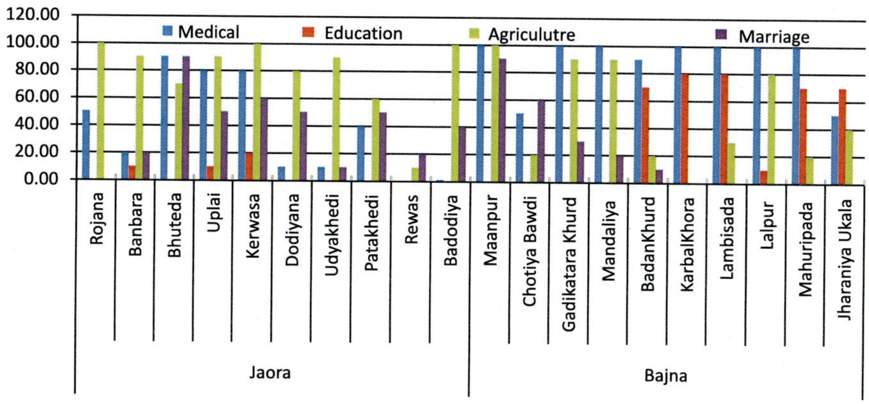

Fig. 5.4.B: Purpose of loan for the households of Jaora and Bajna blocks of Malwa Region.

Source: Primary survey.

Key Issues

1. Higher poverty inhibits money transaction between communities in tribal regions.
2. Availability and access to adequate loans from nationalized banks is lacking in tribal regions.
3. In a tribal region, villagers generally take loans from the local lender and the prime purpose of loan is marriage and agriculture. Further, the accessibility to banks for loans in non tribal regions is better, but due to lesser awareness, villagers usually consult intermediaries/ brokers.

Existing/Ongoing Initiatives

1. Government is promoting Kisan Call Centers (KCCs) at village level. Availability and access to adequate, timely and low cost credit from nationalized banks is prominent in non-tribal regions.
2. Self-Help Groups (SHGs) in non tribal regions are emerging as effective supplementary credit delivery systems for the rural poor including those who are marginally above poverty line and could be termed as "near poor", in particular.

5.1.3. Infrastructure

Infrastructure development is of crucial importance in creating better conditions of life for human societies. Transport and communications infrastructure are important in terms of providing access to basic facilities like health services, education and employment. Further, infrastructures are the fundamental instruments for development. The infrastructural development of the Malwa region was sketched by collecting data on connectivity; accessibility to markets and banks, and availability of government resources such as primary schools and panchayat buildings. Infrastructure is a major component of vulnerability assessment and is composed of four indices and six indicators. The result shows that all the indices are significantly different (Annex 8) in tribal and non tribal blocks. The components (aggregate of indices) however, did not show significant differences between tribal and non tribal blocks because of almost similar programs/policies for development in all the blocks.

Rural connectivity is an important indicator, as it connects people and places through the means of travel and trade and keeps the economy running. Further, it is not only essential for economic development, but it is equally important for social development. In the Malwa region, most of the villages are connected with all weather roads under Pradhan Mantri Sadak Yojana (PMGSY). It aims at providing connectivity to all unconnected habitations with 500 and above general population and a population of 250 in tribal areas with all weather roads. The connectivity index is composed of road connectivity and distance form the nearest bus stop. The connectivity in tribal and non tribal blocks is good while the distance from the nearest bus stop is slightly higher in tribal regions due to remoteness. Moreover, the accessibility to markets for buying or selling agricultural products depends on all weather connectivity. The all weather connectivity results better market accessibility in non-tribal blocks, while market accessibility is lower in tribal blocks (Gandhwani - 0.600 and Bajna - 0.260) due to remoteness (Annex 3 & 5). The task of providing rural connectivity remains one of the most daunting challenges that the state faces today. The Madhya Pradesh Rural Road Development Authority (MPRRDA's) efforts in this direction are remarkable but, at the same time, its operations need to be up-scaled if it were to achieve the set targets within the stipulated timeframe.

Plate-10 a & b : Mode of transportation in tribal villages of Malwa region.

The rural financial system is a powerful instrument and a prerequisite in accelerating the developmental activities in rural areas. Adequate institutional credit is considered to be the most important factor which, if suitably provided, will go a long way to put the economy of the farmers especially the small and marginal farmers on a sound footing (Selvaraj, 1998). The access to banking services is still a daunting challenge in rural Malwa region because of lesser accessibility and awareness. They are forced to travel a long distance for banking and hence the vulnerability score is higher in tribal blocks (Gandhwani - 0.414 and Bajna - 0.375). The government resource index is almost similar in tribal as well as non tribal blocks because of large scale implementation of primary schools and panchayat buildings at the grass root level. Cases of school dropout however, are very high due to a lack of middle and secondary schools. Therefore, efforts should be made to increase the numbers of middle and secondary schools in the rural areas.

Key Issues

- Unavailability of buildings for school and Aangan-wari. Primary schools are running in single rooms.
- Lack of middle and secondary schools, especially in the tribal blocks.
- Maintenance is lacking in schools and panchayat buildings.
- Separate toilets for girl students are still lacking.

Existing/Ongoing Initiatives

- Construction of panchayat bhawan and school buildings.
- Rural connectivity by all weather roads under PMGSY.

5.1.4. Education

The state recorded impressive achievements for education in terms of strengthening its existing school system and innovating for reaching the unreached. Increasing enrolment and ensuring widespread adult literacy in the state have led the effort. As per the Population Census of India 2011, the literacy rate of Madhya Pradesh has shown an improvement of almost 06.39 %. It has reached 70.06% in 2011 from 63.70% in 2001. Further, the efforts of authorities throughout the region to ensure universal access to education is having positive impacts with more and more households able to send their children at least to primary school. There has also been improvement in the gross enrolment ratio. It has risen from 76.5 % in 1996 to 102.33 % in 2007-08. It means that most of the eligible children are enrolled in school. But then the state is still struggling to cope with the school dropout rates. The dropout out rate is 15 % in primary and upper primary levels and a majority of the dropout children belong to SC (17 %) and ST (43.5 %).

a

Plate-11 a : Lower attendance in school of tribal villages.

b

Plate-11 b : Mid day meal significantly improved the attendence in school.

The educational attainment of villagers in the Malwa region was captured with the help of indices such as education index, youth education index; and the final which is impacted by these two indices the decision index. The educational background of the household was measured by the level of education of the household head. The head of the household is of central importance in relation to resource management, strategic planning, and decision making at the household level. The education of the household head is strongly correlated with the literacy of household members as well as the economic wellbeing of the household (Hunzai *et al.*, 2011). Therefore, the educational background proved to be more influential for enhancing capacity of households. The overall literacy is lower in tribal regions than in non tribal regions (Fig. 5.5 A&B).

Box 5.2: Case study of an illiterate woman of Loharpur Village of Gandhwani Block

I never went to school and now it's like being blind. We are poor because we do not have enough land. We don't get money from anything except wage labor. We mainly do wage labor in addition to working in our agricultural field. Mostly, there's no water for agriculture and during rainy season the heavy rainfall damages our crops. The agriculture production is decreasing through the years.

I am an illiterate woman and can't do any business, so I am encouraging my children in their education. I am sure education will bring a bright future. I know the level of education is not appropriate in Government schools but at least they can write and read.

My husband has been working in Gujarat for ten years as a daily wage laborer. He sends Rs. 1000 to 1500 every one to two months. It is never enough for my family consisting of four small children and my husband's second wife and her children. During heavy rainfall, it is very difficult to survive in a non concrete house. We have taken many loans from a local lender. Whatever we earn is either spent on food or to pay back some of the loans. What we get from the agriculture and wage labour is only enough to feed us for a few months.

Access to education, especially in terms of physical infrastructure, is of crucial importance. In both blocks, households have good accessibility to primary education because each panchayat has at least a secondary school; one to two primary schools, and two to three play schools (Aanganwaris). Further, the number of schools in a panchayat depends on the number of villages and populations of that panchayat. But access to education is often affected by the relative remoteness of communities in terms of availability of basic facilities. In remote areas, where access is difficult, even if the physical infrastructure of schools is available, teachers are often unwilling to work there and visit rarely. This constraint severely affects the access to education in many tribal villages.

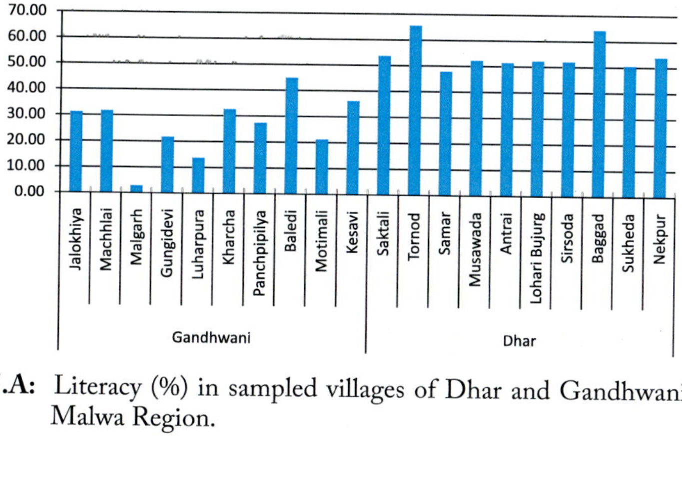

Fig. 5.5.A: Literacy (%) in sampled villages of Dhar and Gandhwani blocks of Malwa Region.

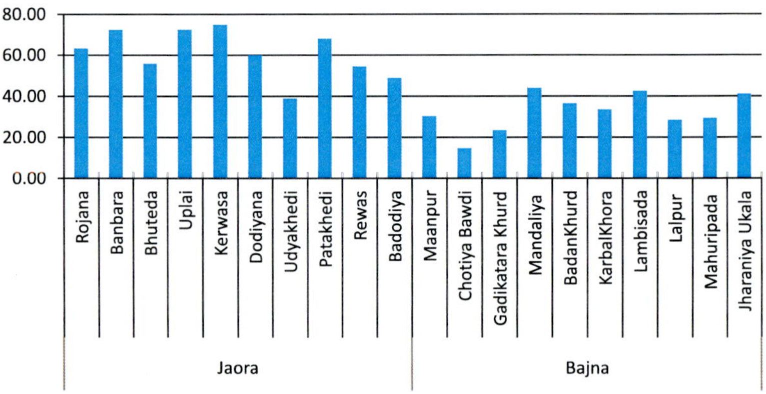

Fig. 5.5.B: Literacy (%) in sampled villages of Jaora and Bajna blocks of Malwa Region.

Source: Census of Madhya Pradesh.

The enrollment and attendance in primary schools of the study area are good and positively impacted by the mid day meal scheme of the government. The day meal allures the children of five to ten years to attend school. But after primary education, most of them give up education because of a number of reasons such as engagement in economic activities, cattle grazing, less inclination toward education, sibling care, social beliefs, parental reluctance and engagement in agriculture. Sometimes, they do not see the need to send their children to school for higher education. Further, the quality of education in the villages is not appropriate and, in most of the cases, a teacher with the help of play school workers runs the school.

The enrolment in high school and higher secondary levels have been reported to be abysmally poor for Madhya Pradesh. Out of the total enrollment (163 lakh children in 2006-07) in elementary education, hardly 15.33 per cent (25.29 lakh) are getting enrolled in high school and higher secondary schools. Further, the proportionate share of SCs and STs is too low. This implies that the secondary education sector in the state needs a fundamental revival to retain the children in the schools. One of the major reasons for this is low access to secondary education in the state. Apart from enrollment, attendance in secondary schools is also a matter of great concern. The attendance in high schools and higher secondary level schools is very low. As discussed, education among the tribes is comparatively very low due to widespread poverty, lack of accessibility and lesser inclination for education. In tribal blocks of the study area, educational vulnerability is higher (Gandhwani - 0.511 and Bajna - 0.543) due to lower literacy among the youth (Gandhwani - 0.46 and Bajna – 0.596), higher proportion of uneducated population (Gandhwani - 0.6760 and Bajna – 0.660), and lower decision making ability (Gandhwani - 0.395 and Bajan - 0.373). On the other side, higher literacy among the youth (Dhar - 0.069 and Jaora - 0.083), lower proportion of uneducated population (Dhar - 0.209 and Jaora – 0.183), and better decision making ability (Dhar - 0.158 and Jaora – 0.186) makes households of non tribal blocks, i.e., Dhar (0.145) and Jaora (0.151) less vulnerable (Annex 2,3,4 & 5). Further a test of significance, using a t-test was done to make sure that there is a statistically significant difference between tribal and non-tribal communities (Annex 8). All the indexes of education were tested statistically (T-test and ANOVA) and they showed a significant difference. The component of education showed significant difference in tribal and non tribal blocks. Apart from this, variability among the tribes of two districts in education status is also measured with the help of ANOVA (Annex 9).

Key Issues

1. School drop-out is higher in tribal regions.
2. Illiteracy among women.
3. Lack of facility for higher education.

Existing/On-going Initiatives

1. Sarba Shiksha Abhiyan seeks to open new schools in those habitations which do not have schooling facilities and strengthen existing school infrastructure through the provision of additional classrooms, toilets, drinking water, maintenance grant and school improvement grants. Under this programme, school buildings are made even in inaccessible pockets of Madhya Pradesh.
2. Aaganbari is a government sponsored child-care and mother-care program running properly in all the sampled villages.
3. Swachh Bharat Swachh Vidyalaya programme is to ensure that every school in India has a set of functioning and well maintained water, sanitation and hygiene facilities. The sanitation facility is enhanced significantly in all the sampled villages and hence it reduced water, hygiene and sanitation related diseases.

5.1.5. Health

The World Health Organization (WHO) claims that "the enjoyment of the highest attainable standard of health is one of the fundamental rights of every human being" (WHO 2006, p 1). The health status of the respondents of Malwa region is measured by two indices, i.e., the health worker index and the health facility accessibility index, and each index is composed of two indicators. The health worker index is composed of frequency of visits of Asha and Doctor in the village while the health facility accessibility index is composed of distance to hospitals and medical shops. The number of medical staff, including doctors and asha worker reflect the capacity of a place or district to cope with health risks. The health worker index showed a slightly higher vulnerability score for tribal blocks (Gandhwani - 0.304 and Bajna - 0.367) as compared to non tribal blocks (Dhar - 0.239 and Jaora - 0.239) due to inaccessibility to health facilities. Apart from visits of asha worker and doctors, proximity to district is also a determining factor for vulnerability assessment. The health facility accessibility index is almost similar for tribal (Gandhwani – 0.517 and Bajna – 0.425) and non tribal (Dhar – 0.470 and Jaora - 0.514) blocks.

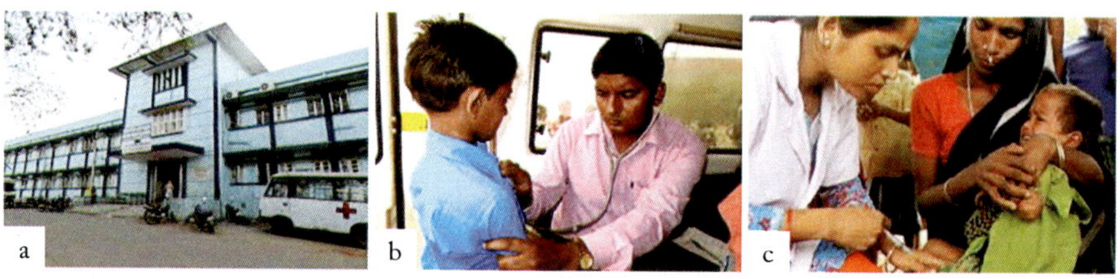

Plate-12 a, b & c : Health check-up and vaccination in Malwa region.

The health status of the region is further validated by asking whether or not households have perceived any new disease during the last five to ten years, and, if so, what kind of diseases have been observed. The cases of diseases coupled with serious illness are commonly observed in the villages of tribal blocks. The higher incidence of diseases might be attributed to poor health facilities and lesser concern for sanitation. Sixty percent of 60% households of non tribal blocks stated that the incidence of diseases increased in the last five to six years and that the occurrence of disease in livestock due to climatic variability is more frequent and commonly observed. The difference in health status of tribal and non tribal villages and among the tribes of the two districts in the same component was tested statistically. The health worker index shows significant difference in tribal and non tribal blocks because of fewer visits of doctors as well as health staff in the tribal villages. Health facility accessibility, however, do not show significant difference in tribal and non tribal blocks because of almost similar accessibility to community health centers and medical shops. The overall component do not show significant difference in health status between tribal and non tribal blocks of Dhar and Ratlam district

Key Issues

- The basic treatment facilities are still lacking in primary health centers.
- Lack of funds for periodic maintenance of hospital buildings.
- Doctor's visit in the hospital is still a big challenge in tribal blocks.
- Inaccessibility to medical shops in tribal villages.

Existing/Ongoing Initiatives

- Primary health center at the panchayat or village level and distribution of free medicine.
- Several state sponsored schemes have been launched for children, women and adolescents like Vijaya Raje Janani Kalyan Beema Yojana; Janani Express Yojana; Janani Sahyogi Yojana; Dhanwantari Block Development Scheme; Prasav Hetu Parivahan Evam Upachar Yojana; Alternate Vaccine Delivery System, and Bal Shakti Yojana. Apart from these other health schemes such as Mother & Child Protection Card; Incentive Scheme for Health Workers (Female & Male) and Anganwadi Workers; Swawlamban Yojana; Prabha Kiran Training Scheme; National Rural Health Mission (NRHM) - Management Information System; New Drug Procurement Policy – 2006, and Mobility Support to Block MO's are also actively promoted.
- Prevention, early detection and appropriate management of micronutrient deficiencies and associated health hazards at the grassroot level.
- Efforts to improve nutritional security of the most vulnerable have improved access to food grains, iodized salt and pulses. Along with this, nutritious food is also provided to adolescent girls and pregnant and lactating women to combat under-nutrition.

5.1.6. Agriculture

Madhya Pradesh has not been amongst the green revolution zones in the 1960s and 1970s, yet it has its own very successful ventures in agriculture that exist with large tracts of traditional low technology and dry-land farming areas. With very little industrialization and lack of developed sectors, the state showed higher dependence on the primary sector. The primary sector provides employment to 70% of the state's population, but has had very wide fluctuations across the last ten years due to dependence of agriculture on the monsoons. The two-third gross cropped area of the state depends on rain and almost 80% area is vulnerable to frequent natural calamities. The nature of land holding also affects agriculture of the state. It is recorded that nearly three-fourth holding of the state is possessed by small and marginal farmers. The larger tribal population, along with a higher proportion of marginal and small farmers, is the principal constraint for adoption of high technology in agriculture. The constraints in agriculture are further compounded by a weak rural credit system, less diversified agriculture, higher proportion of indigenous livestock, and fodder shortage in spite of a large proportion of fallows/wasteland and labour intensiveness.

Plate-13 a : Traditional agriculture in Malwa region.

Plate-13 b : Agriculture of tribal villages.

Plate-13 c : Agro-forestry practice in non tribal villages.

As discussed, 72 % of the workforce is engaged in agriculture, but its contribution to the GSDP is declining continuously since 1993-94 (36 % - 1993-94; 31 % - 1999-2000; 25.24% - 2005-06; 23.88 % - 2006-07; 22.28% - 2007-08; 21.71% - 2009-10 and 20.40% - 2010-11). After 2010, however, the availability of quality seed and access to irrigation facilities improved significantly, resulting in enhanced production/hectare (1064 kg/hectare - 2007-08 to 1952 kg/hectare - 2012-13). Thus the contribution of agriculture to the GSDP has increased (22.28 % - 2011-12; 24.35%– 2012-2013 and 27.38%– 2013-14) considerably and, therefore, the sector becomes dominant for determining economic performance.

The major challenge in the agricultural sector is to improve the poor quality of agriculture in many areas. Agriculture of the Malwa region especially the tribal blocks is challenging; it holds the key to livelihood security and rapid reduction in rural poverty. It is expected that the ongoing population growth, urbanization, and changes in climate will further reduce the availability of water resources (Ragab and Prudhomme, 2002; Molden, 2007; Hanjra and Qureshi, 2010) which will affect agriculture. In Madhya Pradesh, climate change is likely to intensify and increase the occurrence of drought. Coupled with rising temperatures, soil moisture levels will be placed under more stress. These factors will exacerbate existing pressures on rural livelihoods. As with food security, the households that are already facing water shortages are more sensitive to changes in temperature and precipitation.

The agriculture of the block is further constrained by the undulating topography and intensiveness. The present status of agriculture of the Malwa block is captured with the help of seven indices and 13 indicators, i.e., agricultural equipment, land, cropping intensity, yield, irrigation and cropping strategies. All of the indices showed significant differences between tribal and non tribal blocks, except the agricultural strategy index, that could be attributed to similar strategies. Further, the component is also tested statistically and a significant difference was observed between tribal and non tribal blocks.

Plate-14 a & b : Agro-horticulture practices in Malwa region.

Basic agricultural equipments like tractors, power tillers and motor pumps are of critical importance for agriculture. It is observed during repeated interaction with villagers that the households of non tribal blocks were well equipped for agriculture. In these blocks, most of the households possess tractors, power tillers and moter pumps for agricultural purposes but the agricultural equipment are lacking in tribal blocks. Therefore, the agricultural equipment index showed higher vulnerability in tribal blocks (Bajna - 0.940 and Gandhwani - 0.756) and lower vulnerability in non tribal blocks (Jaora - 0.171 and Dhar - 0.232).

The average landholding size of the state is 2.2 hectare. Thus, the majority of the farmers come in the category of marginal and small. They mostly practise subsistence farming and do not have adequate resources to invest in their land. The same trend is also observed in Dhar and Jaora blocks of the Malwa region. The average land holding varies between 2.5 to 03 hectare, out of which about 80-90% land was irrigated (Fig. 5.6.A&B). The agriculture of the region is profitable because of accessibility to better facilities especially irrigation. On the other side, average land holding in the Gandhwani and Bajna blocks varies between 0.1 to 1.5 hectare and only 25-30% land was irrigated. The lower landholdings, together with lesser irrigated areas of tribal blocks, make them vulnerable in terms of agricultural land (Gandhwani - 0.712 and Bajna - 0.829).

Madhya Pradesh, one of the largest states in the country, suffers from inadequate availability of irrigation water, mainly because of its undulating topography, high surface runoff, and lack of surface storage facilities. A greater part of the state, i.e., 63.50% is rainfed and only 36.50% area is irrigated. The gross cropped area is 217.55 lakh ha with cropping intensity of 144.87%, and the net sown area is 150.17 lakh ha constituting 49% of the total geographical area (2011-12) (SFP, 2014-15). The unavailability of irrigation facilities affects the cropping intensity and production. The better

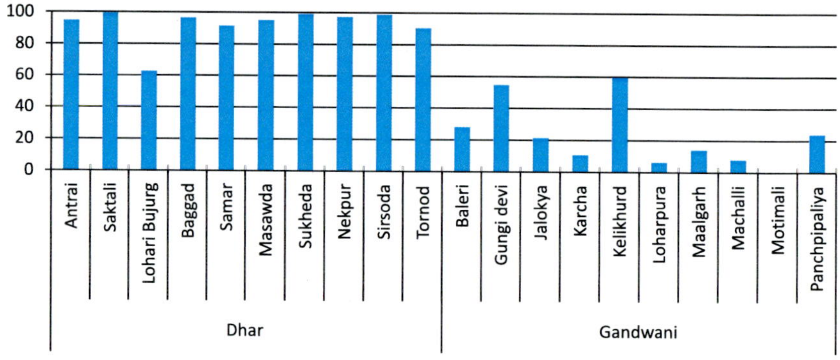

Fig. 5.6.A: Percentage of irrigated areas in Dhar and Gandhwani blocks of Malwa Region.

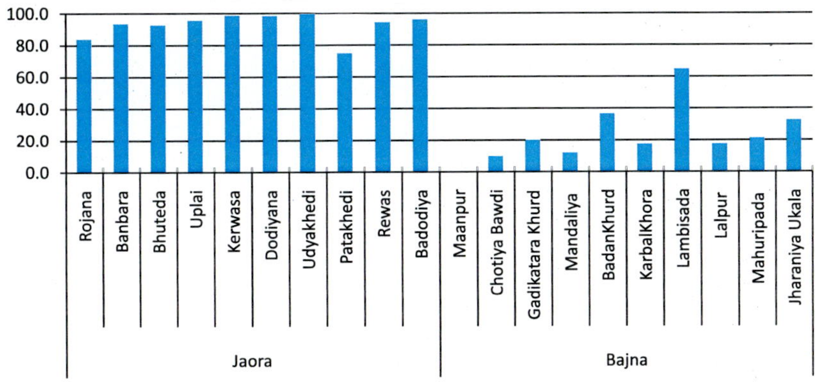

Fig. 5.6.B: Percentage of irrigated areas in Jaora and Bajna blocks of Malwa Region.

availability of irrigation facilities reduces the vulnerability of Dhar (0.200) and Jaora (0.174) blocks, while unavailability of irrigation facilities augment the vulnerability of tribal (Gandhwani - 0.709 and Bajna - 0.691) blocks (Fig. 5.7. A&B). In these block, due to lesser water availability, land is often kept fallow in the Rabi season and crop taken only in the Kharif season. Mono cropping increases the risk of yield loss from extreme weather events and changes in temperature and precipitation (Abramovitz *et al.*, 2001), and thus raises the sensitivity of households. On the other side, cropping intensity was almost double in non tribal blocks due to better irrigation availability. Most of the households of this block take two crops in a year, i.e., Rabi - wheat and Kharif - soyabean and maize. Therefore, the higher cropping intensity reduced the agricultural vulnerability of the non tribal blocks.

Non-irrigated agriculture depends entirely on rainfall. This form of agriculture is possible only in regions where rainfall distribution ensures continuing availability of soil moisture during the critical growing periods for the crops. The situation in the Malwa region is slightly different, soil moisture is low because of low rainfall, and this is one of the major constraints for agriculture. Further, the recurrent drought and rainfall affects agriculture almost every alternate year. The higher reliance on agriculture for livelihood makes them more vulnerable.

In tribal blocks, agriculture is totally based on rain and access to irrigation was confined to a few villages due to poverty, low ground water level and

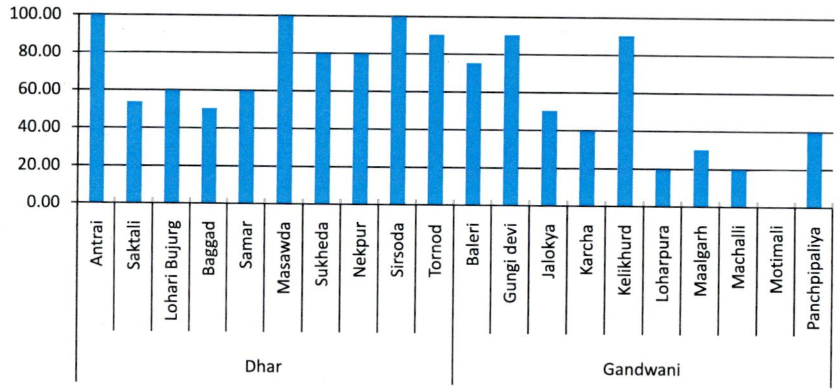

Fig. 5.7.A: Percentage of households having irrigation facilities in Dhar and Gandhwani blocks of Malwa Region.

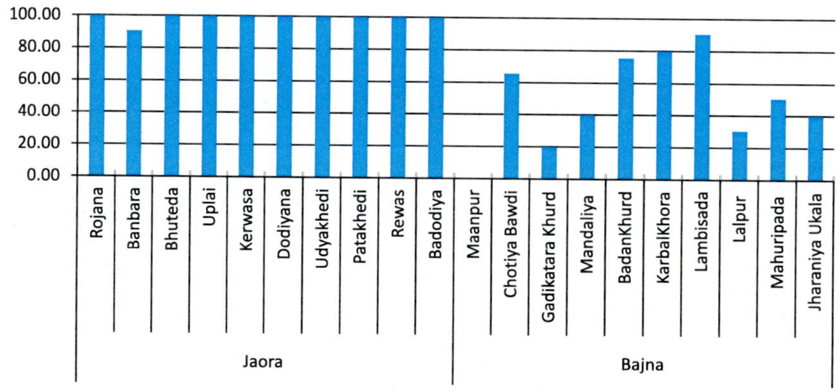

Fig. 5.7.B: Percentage of households having irrigation facilities in Jaora and Bajna blocks of Malwa Region.

lesser implementation of structures on water conservation. On the other side, irrigation accessibility was comparatively better in the villages of non tribal blocks. Most of the villagers' possess dug well or tube-well for irrigation. Further, the availability of irrigation in tribal blocks varies between 02 to 03 months after rainy season while it 03 to 04 months for non tribal block (Fig. 5.8. A&B). The major source of irrigation in the state is wells and tube wells (65%), canals (17%) and others (17%) (SFP, 2014-15). Similarly, in Dhar and Ratlam districts, well and tube wells are the major source of irrigation. Canals, natural irrigation sources and rivers, however, are also used for irrigation purposes (Fig. 5.9. A&B).

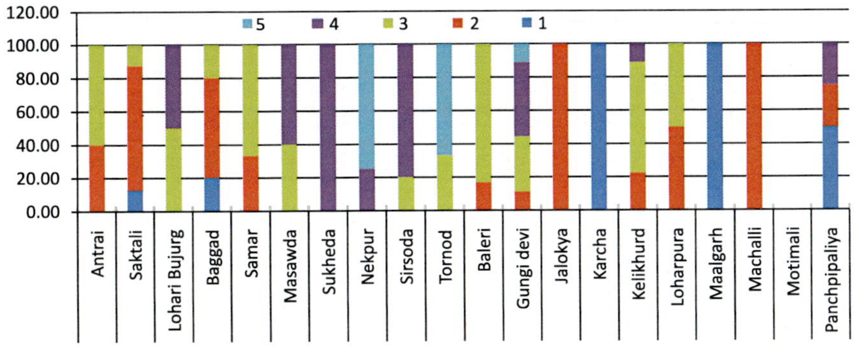

Fig. 5.8.A: Water availability (month) for irrigation of Dhar and Gandhwani blocks of Malwa Region.

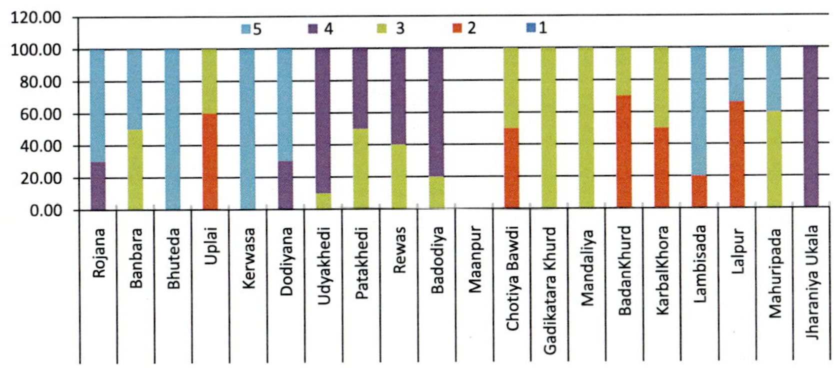

Fig. 5.8.B: Water availability (month) for irrigation in Jaora and Bajna blocks of Malwa Region.

The number of irrigation source available to Rabi crop depends on availability of water in the water bodies. The water availability for longer periods results in more number of irrigation sources to Rabi crop and generally the yield of Rabi crop depends on the number of irrigation sources. For instance, wheat requires four to five irrigation sources for proper yield (Fig. 5.10. A&B). Lesser irrigation accessibility and availability affect the agriculture and livelihood of villagers and make them more vulnerable. Therefore, the vulnerability of tribal blocks (Gandhwani - 0.523 and Bajna - 0.486) is higher than in non tribal blocks (Dhar - 0.252 and Jaora - 0.146) due to non-navailability of irrigation (Annex. 2,3,4 & 5).

The irrigation strategy in both the blocks is not appropriate. In non tribal blocks, farmers change the crop type, irrigation pattern and a few of them apply for government support and, therefore, the vulnerability is lower (Dhar - 0.340 and Jaora - 0.325). In addition to change in crop types, proximity to district and accessibility to technical support for agriculture is also considerable factor for better strategies. Agricultural strategies are lacking in tribal blocks making them more vulnerable (Gandhwani - 0.495 and Bajna - 0.497).

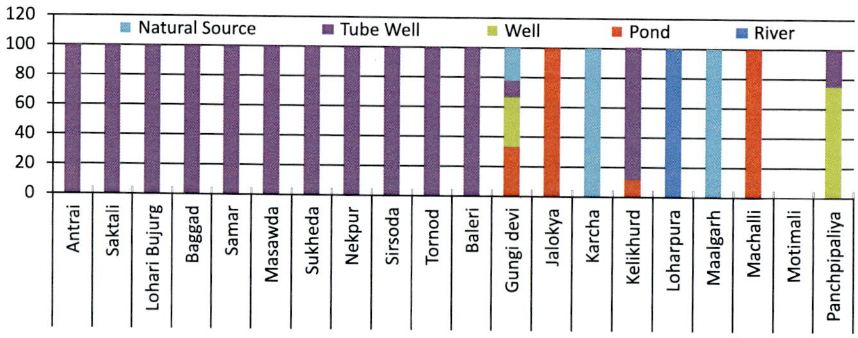

Fig. 5.9.A: Sources of irrigation in Dhar and Gandhwani blocks of Malwa Region.

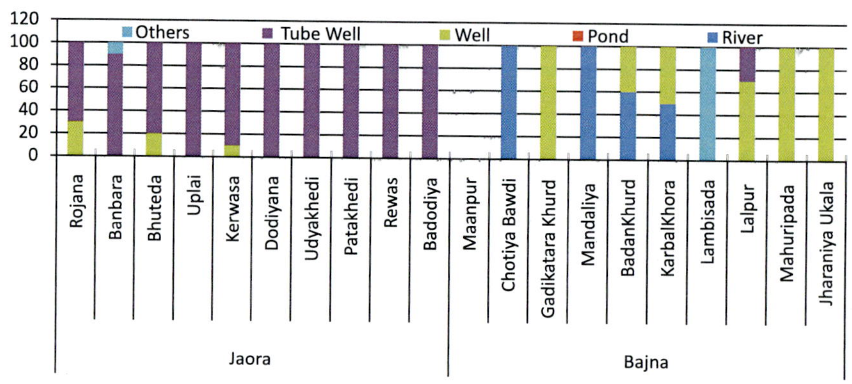

Fig. 5.9.B: Sources of irrigation in Jaora and Bajna blocks of Malwa Region.

There is widespread consensus that variations in temperature, precipitation, and extreme weather events affect livelihood. There are direct effects on health from heat and cold waves as well as indirect effects via food production and nutrition, vector-borne infectious diseases, water-borne infectious diseases, and social and economic disruption (Githeko and Woodward, 2003). To take this into account, a section is framed in the questionnaire on the perception of people on summer temperature (variation in summer

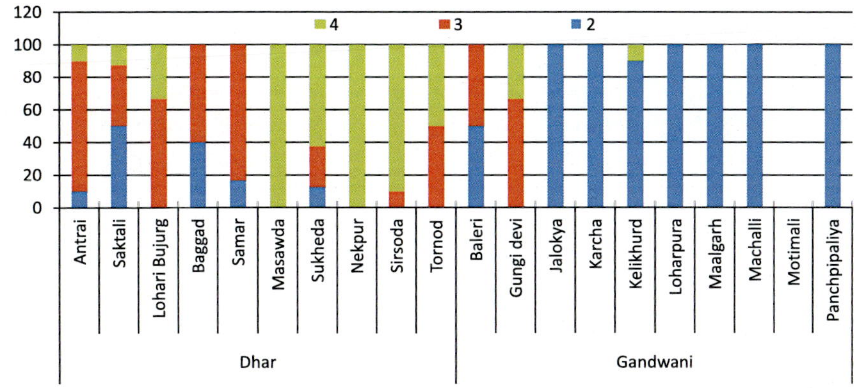

Fig. 5.10.A: Number of different irrigation sources in Dhar and Gandhwani blocks of Malwa Region.

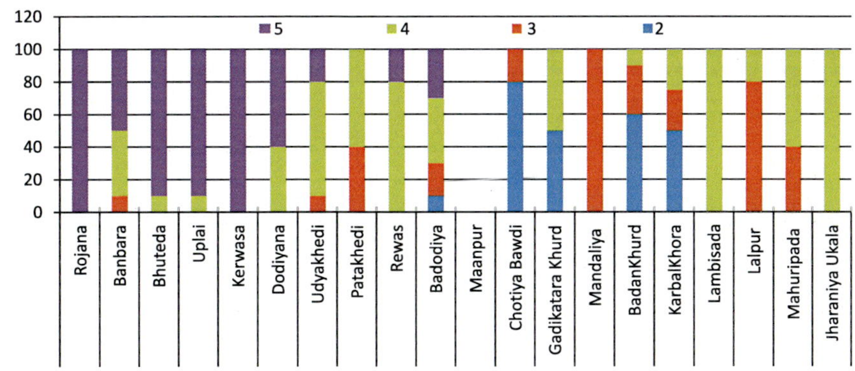

Fig. 5.10.B: Number of different irrigation sources in Jaora and Bajna blocks of Malwa Region.

temperature and frequency of hot months), rainfall (frequency, intensity and rainfall pattern) and winter (variation in winter temperature). It was generally observed through repeated interactions with the villagers of Gandhwani block that they were living in proximity to nature and were affected more by climatic variations. The higher exposure in terms of change in temperature and rainfall directly affects the livelihood of villagers and make them more vulnerable.

Natural calamities are measured by the perceived changes in climate events and in the occurrence of extreme weather events and hazards during the last ten years. The concept covers changes in the frequency and severity of extreme events. To assess the natural calamities, indicators like change in frequency and intensity of drought, hailstorm and cold wave were used, a central one being the number of extreme events during the last ten years that a household had experienced. The natural calamities index of both blocks depicted more or less similar vulnerability. The agricultural strategies are slightly better in Dhar district (Dhar - 0.103 and Gandhwani - 0.112) as compared to Ratlam district (Jaora - 0.206 and Bajna - 0.186). The higher vulnerability of Ratlam district might be attributed to lesser inclination for the formulation of strategies. Further, extreme events put a lot of stresses on households of the Malwa region and it was found that climatic stress is higher in tribal blocks due to their higher exposure and lower adaptive capacity.

Key Issues

1. Limited area under irrigation due to undulating terrain and inadequate infrastructural development.
2. Subsidy and support during extreme events, especially during crop damage, are lacking in tribal regions.
3. Continued vicious cycle of low agricultural production, especially in tribal regions, due to unawareness of new technologies and lack of extension services.
4. Un-recommended dose of pesticides and fertilizers leading to depletion of humus and nutrients.

Existing/Ongoing Initiatives

1. Realizing optimum irrigation potential under major, medium and minor irrigation projects.
2. Conserving and protecting ground and surface water bodies by constructing engineering structures.
3. Subsidy on improved and technologically feasible irrigation systems such as drip and sprinkler irrigation systems.

5.1.7. Food

Madhya Pradesh has topped in food grain production in the country with a food grain production of 277 lakh tons in 2012-13, including 161 lakh tons of wheat. The contribution of Madhya Pradesh to India's total food grain and wheat production is 11.2% and 17.5 % respectively during 2012-13. The impressive performance of the state in the agriculture sector is attributable to increase in irrigation capacity, availability of quality seed and fertilizer, and improved infrastructure. As a result of higher production, the state implemented the National Food Security Bill with the objective of providing food and nutritional security to people. The government ensures the guaranteed supply of food grain at highly subsidized rates to the beneficiaries. Climatic variability along with poor quality agriculture, however, has kept many areas in the economic scenario in a depressing position.

Recent projections show that climate change will have an impact on food security (food production, stability and access to food) (Gregory *et al.*, 2005; Schmidhuber and Tubiello 2007). "The importance of the various dimensions and the overall impact of climate change on food security will differ across regions and over time and, most importantly, will depend on the overall socio-economic status that a country has accomplished as the effects of climate change set in" (Schmidhuber and Tubiello, 2007). At the micro level, those households that are already unable to maintain food security are the ones that are most sensitive to climate variability (Maxwell and Smith, 1992). To assess the food security of households, two indices and three indicators were used, a central one being the primary source of food, food sufficiency to feed all household members throughout the year, and variation in consumption of nutritious food in the last ten years. The food sufficiency index is significantly different in tribal and non tribal blocks while the nutritious food index did not show any significant difference. The overall components also do not show significant differences between tribal and non tribal blocks. Further, the difference in food sufficiency and availability among the tribes of two districts in the same component was tested with the help of ANOVA.

Agriculture is the prime source of food in the Malwa region, although a few of them also depend on the PDS (Fig. 5.11. A&B). The vulnerability score for sources of food and their availability throughout the year is higher for Gandhwani (0.270) and Bajna (0.235) blocks. The higher vulnerability score can be attributed to low agricultural production, coupled with lower landholdings. The agricultural production of the block is generally affected by inaccessibility to basic facilities for agriculture like irrigation, improved seeds and modern technology. The agriculture of the blocks is also labour intensive and often affected by climatic variability. This is turn, affects the availability of sufficient food. Since 2011, the agricultural production of Gandhwani block has continued to decline due to climatic variability.

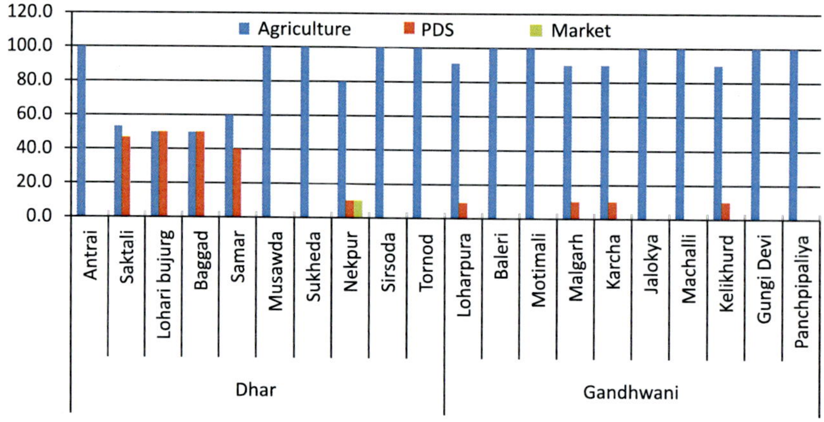

Fig. 5.11.A: Percentage of dependency on agriculture, PDS ad market of households of Dhar and Gandhwani blocks of Malwa Region.

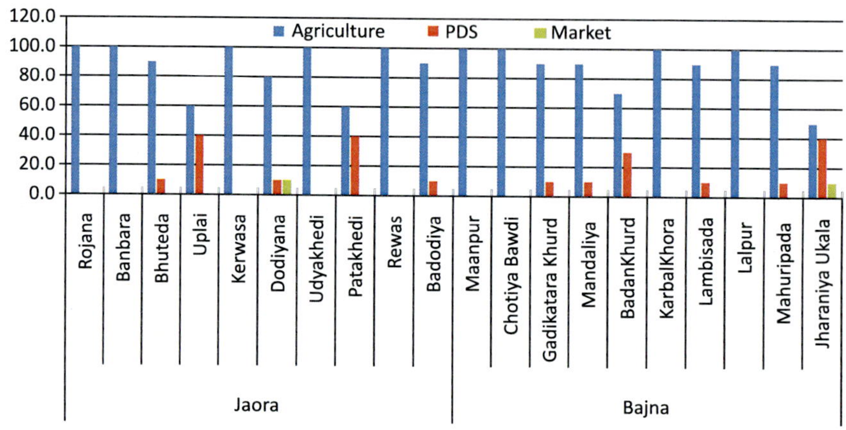

Fig. 5.11.B: Percentage of dependency on agriculture, PDS ad market of respondents of Jaora and Bajna blocks of Malwa Region.

Source: Primary survey.

In the Gandhwani block, only 16% of households get sufficient food for nine to twelve months, 27% for six to nine months, 55% for three to six months and the remaining 2% get sufficient food for zero to three months from agriculture (Fig. 5.12. A&B). On the other side, 80% of households of the Dhar block get sufficient food for nine to twelve months and the remaining get sufficient food for six to nine months. Similarly, only 8 % households of Bajna block get sufficient food for nine to twelve months; 20% for six to nine months, 50% for six to three months and the remaining 22% get sufficient for zero to three months from agriculture. On the other side, 76% of the households of the Dhar block get sufficient food for nine to twelve months and the remaining get sufficient food for six to nine months from agriculture. This clearly indicates that agricultural production of non tribal blocks is comparatively better than tribal blocks and satisfies the need of a large section of the population for a longer span of time. The higher food availability can be attributed to comparatively better agricultural knowledge; and awareness and proximity to the district headquarters. Therefore, vulnerability of Dhar (0.219) and Jaora (0.147) blocks for sources of food and their availability throughout the year is low.

Adults in Madhya Pradesh suffer from a dual burden of malnutrition; two in five adult women and men (42 %) are too thin, and 08 % of women and 4 % of men are overweight or obese. Only 51 % of women and 54 % of men are at a healthy weight for their height. Under-nutrition is much higher and is a serious issue for the poor and especially for scheduled tribes and scheduled castes. Further, the per capita expenditure on food is the lowest in Madhya Pradesh. This is primarily a manifestation of widespread poverty and lack of livelihood security, and is arguably the most important factor responsible

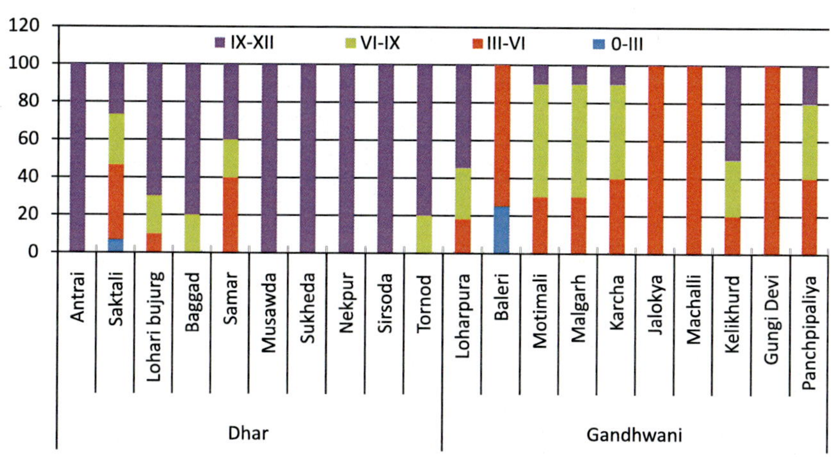

Fig. 5.12.A: Availability of food (months) from agriculture for the respondents of Dhar and Gandhwani blocks of Malwa Region.

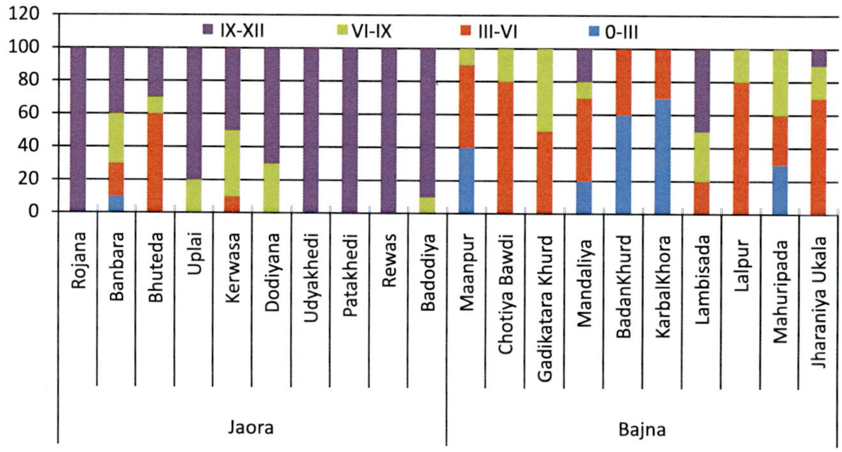

Fig. 5.12.B: Availability of food (months) from agriculture for the respondents of Jaora and Bajna blocks of Malwa Region.

Source: Primary survey.

for determining the poor nutritional status of people in the state. A similar trend is observed in the Malwa region, where the households of Dhar and Jaora blocks reported that consumption of nutritive food is decreasing in the last ten years, although agriculture of the non tribal blocks is diversified and production is comparatively better. Along with this, households were well calculative for other needs of daily life and spent a small portion of their income on food. On the other side, households of tribal blocks invest a large portion of income on food. Earlier, they rarely used to eat nutritive food like pulses, vegetables, meat and milk but now nutritive foods are in their diet. Therefore, non tribal blocks, especially the Dhar block showed slightly higher vulnerability for change in consumption of nutritious food.

Key Issues

1. Higher dependency on agriculture for food and very few of them getting sufficient food throughout the year in tribal block.
2. Lack of availability of nutritious food.

Existing/Ongoing Initiatives

1. The Public Distribution System of Madhya Pradesh facilitates the supply of food grains and distribution of essential commodities to a large number of poor people through its network of Fair Price Shops at a subsidized price on a recurring basis.

5.1.8. Water

Water security is of vital importance to the livelihoods of people in the Malwa block. It is expected that population growth, urbanization, and changes in climate will further reduce the availability of water (Ragab and Prudhomme, 2002; Molden, 2007; Hanjra and Qureshi, 2010). As with food security, the households that are already facing water shortages are more sensitive to changes in temperature and precipitation. The sensitivity of households further increased due to inaccessibility to potable water. Therefore, accessibility together with availability of potable water is of vital importance and should be addressed during vulnerability assessment. Three indices and eight indicators on potable water including source, availability and accessibility of water, time spent, and distance travelled for water collection were used to measure water vulnerability of the Malwa region. The water accessibility and availability indices do not show significant differences but the water collection index showed significant differences between tribal and non tribal blocks. Further, the component showed significant differences between tribal and non tribal blocks. The difference in water accessibility, availability and collection within the tribes of Dhar and Ratlam district was also test statistically and presented in Annex 8.

In Madhya Pradesh, about 68.4 % of households have access to safe drinking water and out of the total, 86.4 % are dependent on groundwater in which hand pumps and tube wells constitute 50.8 % and wells contribute 35.6 %. Tap, river/canal, tank and others contribute only 13.6 % of the total water supply (Das, 2004). Similarly, hand pumps are used as a source of drinking water in the Malwa region, but during summer, they also use tube-wells. Expect for a few, all villages have one or two sources of water inside the village boundary, but the higher dependency and lower ground water level limit the water availability from eight to nine months. The villagers suffered from huge water scarcity in the remaining months of the year, especially during summer.

Field visits and interviews with villagers have shown that the majority of households lack basic amenities like running water for drinking and Field visits and interviews with villagers have shown that the majority of households lack basic amenities like running water for drinking and household purposes, sanitation, and electricity. The non-availability of water for drinking and other purposes is considered a major problem (Fig

Plate-15 a, b & c : Sources of drinking water.

5.13 A& B). This problem is further compounded due to the lesser number of water sources and higher dependency on them. In many villages, whole populations were dependent on single sources of water throughout the year. During times of technical faults in sources or in summer, they collect water from other villages. The situation worsens during drought as some of them have to travel up to four kilometers. In addition, the geographical location of the villages made it difficult for locating new water bodies because many of them were sparsely distributed in remote areas.

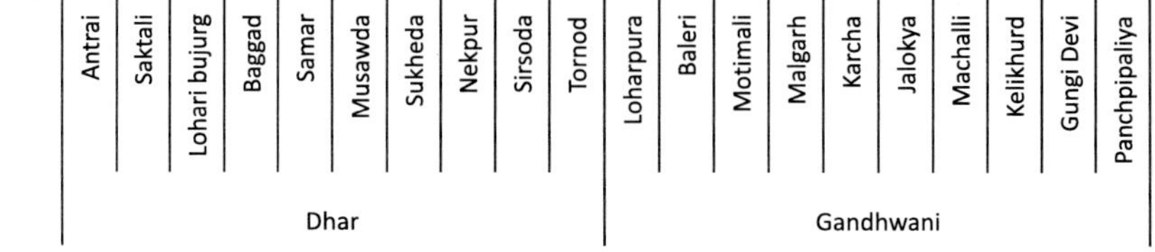

Fig. 5.13.A: Availability of drinking water (months) for the selected respondents of Dhar and Gandhwani blocks of Malwa Region.

The complexities for non tribal blocks of the Malwa region was comparatively low due to good infrastructural development, proximity to the district head-quarters and higher awareness among villagers. Other water related issues like excess water withdrawal for agriculture, however, was observed in some villages. Villagers generally use borewells/tubewells and wells (water lifting through motor pumps) for irrigation purposes. After irrigation, most of the water bodies (wells) whose water level was low, dried due to excess withdrawal of water for irrigation purposes. This creates a water conflict amongst villagers and affects the sources of drinking water.

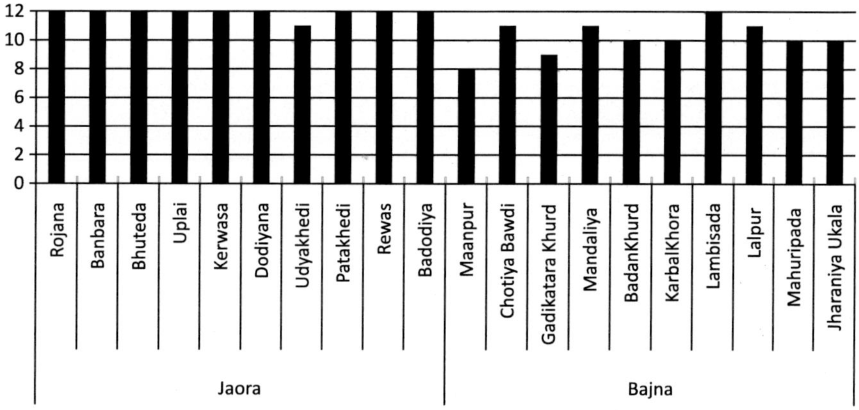

Fig. 5.13.B: Availability of drinking water (months) for the selected respondents of Jaora and Bajna blocks of Malwa Region.

Source: Primary survey.

The vulnerability index for water accessibility constitutes indicators like sources of potable water and their alternatives during lean seasons. The vulnerability score for source of potable water is higher for the tribal blocks (Gandhwani - 0.368 and Bajna - 0.261). This could be attributed to lesser number of sources with higher dependency on them. Again, the vulnerability due to alternatives for drinking water, or sources of water during the lean season, was approximately the same for blocks in Dhar district (0.131 - Gandhwani and 0.142 - Dhar). A slight variation in the vulnerability score of Ratlam district (0.222 - Bajna and 0.102 - Jaora) is also observed. The lesser variation in vulnerability scores are credited to similar geographic characteristics of the region. Moreover, the region is known as a water scare zone of Madhya Pradesh and villagers of entire blocks have to travel more or less the same distance for water collection during lean seasons.

It was generally observed that the average distance travelled for water collection in non tribal blocks, is 200 meters and takes about 25 minutes, and this is increased to 600 meter and an hour and a quarter during summer.

The percentage increase in distance and time for collection of water during summer is 200 and 140 respectively. On the other side the average distance for water collection in tribal blocks is 300 meters and it takes about 55 minutes during summer the distance is further increased to 1100 meters and the time is increased to 02 hours and 20 minutes (the percentage increase for distance and time is 520 and 220 respectively). Therefore, the vulnerability scores for collection of water is very high for tribal blocks, i.e., Gandhwani (0.729) and Bajna (0.432), as compared to non tribal blocks, i.e., Dhar (0.312) and Jaora (0.073).

Key Issues

1. Over extraction of water leads to ground water table decreasing.
2. Lack of maintenance of existing sources of water. In tribal regions, villagers were dependent on a single source of water and more than 50% of hand-pumps were not working due to technical faults.
3. Ground water recharge is lacking in all the sampled villages.
4. Water shortage is prevalent during summer, especially in tribal regions.

Existing/Ongoing Initiatives

1. Nal Jal Yojana is to provide potable piped water supply to rural households. A majority of non tribal villages benefits from the scheme but it is lacking in tribal regions.
2. Drinking water supply through hand pumps in water scarce regions.
3. Water supply is provided to the villagers with the help of a power pump, pipeline and cistern.
4. Water supply schemes in tribal areas (as notified by the Govt. of Madhya Pradesh) by providing either hand pumps or piped water supply schemes.
5. Water storage tanks and facility of tankers in water scarce regions under the Vidhayak Vikash Nidhi.
6. Integrated and co-ordinated efforts by different block level agencies for ground water recharge such as MGNREGA, Integrated Watershed Management Program (IWMP), forest department, etc.

5.1.9. *Weather Perception and Calamities*

The impacts of climate change are expected to be higher for some segments of society because of geographic location, the degree of association with climate-sensitive environments, and the unique economic, political, and cultural characteristics of particular areas (McCarthy *et al.*, 2001; Dolan and Walker, 2004). The vulnerability of climate sensitive sectors and associated communities is comparatively high because the changes may affect their activities and values.

The weather prediction and calamities index was designed to capture the perception or views of rural communities on climatic variability and change. The level of education, awareness, access to weather forecast, reliability of forecast, involvement of formal/non-formal institutions, and better communication are critical factors for assessing vulnerability and formulating coping strategies. The weather perception and calamities index is composed of six indices and 15 indicators. Except for the natural calamities index, all the indices showed significant difference in tribal and non tribal blocks. In totality, the component of weather perception and calamities showed a significant difference between tribal and non tribal blocks. The perception on climate change showed significant differences within the different tribes and non tribes of Dhar and Ratlam district, presented in Annex 8.

The limited access to weather forecast in terms of radio, TV, newspaper, and panchang makes the households of tribal blocks (Gandhwani - 0.377 and Bajna - 0.411) more vulnerable. On the other side, households of non tribal blocks (Dhar - 0.328 and Jaora - 0.317) had better accessibility to weather forecasts due to awareness, better communication and strategic settings. Further, the households of tribal blocks rely on traditional forecast systems and have better traditional knowledge (Box. 5.1.). The reliability of weather prediction through the traditional knowledge system is still better, but it is not transferred properly to the next generation. Even though the reliability of forecast of damage due to extreme events is higher in tribal blocks because of lesser coping strategies against climate variation. In non tribal blocks households having good accessibility to weather forecast, but the use of traditional knowledge has vanished and hence the losses during extreme events is higher. Therefore, integration of scientific techniques and traditional knowledge in climate science is required for strengthening the weather forecast system.

Box.5.1: Traditional Weather Forecast Practices

Indigenous knowledge is a systematic body of knowledge acquired by the local people through accumulation of experience, informal experiments and an intimate understanding of the environment in a given culture (Anandraja *et al.,* 2006). Indigenous people have developed a rich knowledge base of predicting weather events. This system is helpful in minimizing the threats of extreme events and making coping strategies. The indigenous knowledge system of the Malwa block is discussed below:

- Location and pattern of clouds is a good indicator of rain. Dark, dense clouds indicate early and heavy rainfall. In case dogs cry all night, rain may get delayed.
- Incessant calling of pheasant/cricket in the morning and the relentless singing of the cuckoo indicates immediate onset of rain. If the pheasant continues calling during rainfall, it signifies that the rain will stop soon. Further, the intensity and duration of rain is predicted by the movement of hen/cock in search of food. Their continuous movement indicates intensified and longer duration of rain. If they stop searching for food, it implies that the rain will stop soon or will be less.
- Rain is also predicted through the movement of ants. If group of ants moves along a definite path carrying their food, heavy rain is expected on the same day or within two days.
- When frogs croak in the afternoon and continue till evening, it indicates the onset of monsoon.
- Moisture in dried chilly is a good indicator of heavy rain.
- Prediction of rain is also based on the onset of monsoon in India. In the Malwa block, monsoon generally begins after ten days of the onset of monsoon in Mumbai.
- When the trees shed their leaves and the fruits of the *neem* ripen and drop, it indicates the onset of monsoon.
- Frequent appearance of the rainbow indicates rain except if it is seen in the east.

Key Issues

1. Lack of forecasting facilities.
2. Traditional knowledge system is deteriorating.

Existing/Ongoing Initiatives

1. Regular training for villagers.
2. Weather forecast through "Krishi Vigyan Kendra".

5.1.10. Natural Resources

Natural resources are of paramount importance not only from the economic point of view, but also from the ecological point of view. It has a multifarious role in maintaining the ecological balance and moderation of the environment, besides providing livelihood opportunities to people depending on forests. From ancient times, forests and people have had a symbiotic relationship due to the peculiar characteristics of forests as an economic resource, as a legacy and a part of the future. The growing human population and consequent large scale urbanization, industrial development and political consolidation have put a heavy demand on forests for fuel, fodder, food, medicine and small timber causing degradation of forests. According to FSI (2011), the very dense forest of the state in the last two years (2009 to 2011) decreased by 7 sq. km., moderately dense forest by 21 sq. km., scrub forest by 5 sq. km. and the area under open forest increased by 28 sq. km.

Box 5.2: Traditional Folklore for Rain

Tribals have a rich culture in the form of social organizations, laws, festivals, music, dance, drama, etc. They have rich cultural practices and traditional knowledge like the practice of Halma, vanpuja etc.

1. Halma

Halma is the tribal tradition under which a person in distress can seek help from his fellow tribals. For example, if a tribal household wants to make house but is unable to invest in the required daily wage labour, then under *Halma,* households of the community together and support in different ways. They work in small groups, some bring wood from the forest, and some start manual labor work to help make the house. Earlier, tribes mostly called *halma* for conserving water to satiate the thrust of mother earth.

2. Devi Maa Ka Jungle

The practice of *Vanpuja* or *Devi Maa ka Jungle* involves preserving the sacred patches of forest called '*Van*' where the forest goddess is worshipped and tribals do not cut the trees around that place, they do not even use the waste wood from there.

In the Malwa region, livelihoods of a large section of the population depends on forests. About 80% of the selected households collects fodder and fuel wood from the forest, but the actual issues are: do they get sufficient fodder and fuel-wood from the nearby forest? how much time do they spend on collection of fodder and fuel-wood? is the distance between forest and village increasing? is the quality of natural resources deteriorating; and what is their strategy to cope with climatic variability. Questions on the availability of fodder and fuel-wood and their sufficiency were also included in the questionnaire. The distance of forest, time spent for fodder and fuel-wood collection, quality of natural resources, and their coping strategies are assessed and addressed in this section. There are six indices, and ten indicators were selected for assessing natural resource vulnerability of the Malwa region. All the indices were tested statistically. Natural resource collection and degradation, as well as natural water indices, do not show significant differences. Natural resource availability, fodder and fuelwood indices, however showed significant differences between tribal and non-tribal blocks. Further, the overall component was also tested for analyzing differences between tribal and non tribal blocks and the component showed a significant difference. The perception on climate change also showed a significant difference within the different tribes of Dhar and Ratlam districts, as presented in Annex 8.

Plate-16 a, b, c & d: Forest and plantation on Malwa region.

Natural resource availability indices are calculated by using two indicators, i.e., percentage of households that collect fodder and fuel-wood from the forest and percentage of households that do not get sufficient fodder and fuel-wood. The forest cover in the Malwa region is deteriorating at an alarming rate due to higher dependency. Hence the vulnerability of natural resources varies according to dependency and availability of sufficient forest products. The vulnerability score for availability of natural resources is contracting in the first tribal block, i.e., Gandhwani, where the forest cover is continuously decreasing due to unsustainable harvest and lack of management practices. About 90% respondents solely depend on the forest and only 48% gets sufficient forest products and hence vulnerability in the Gandhwani (01.00) block is higher. Vulnerability is however, lower in the second tribal block, i.e., Bajna (0.380) because of controlled grazing and sustainable management, including afforestation. Moreover, 90% of the households solely depend on the forest for forest products, out of which 78% get sufficient forest products from nearby forests. About 05% and 35% households in Jaora and Dhar blocks respectively were dependent on forest for forest products and not a single household gets sufficient quantity of forest products. However, the per capita forest area in the state (0.16 ha) is much higher than the national average (0.07 ha) (FSI, 2011).

The natural resource collection index is composed of distance travelled and time spent for collection of forest products, i.e., fodder and fuel-wood. The distance travelled and time spent for collection of forest products is approximately the same in Ratlam and Dhar districts. In Dhar district a higher proportion of the population was dependent on limited forest resources leading to a higher rate of forest degradation and distance between forest and village is increasing continously. The distance between forest and villages, and time spent for collection of forest products is approximately the same and hence the vulnerability score of both blocks of Dhar district is the same, i.e., 01.00. On the other side, in Ratlam district, the dependency on forest is comparatively low due to access to other alternatives. Therefore, the vulnerability score in the blocks of Ratlam (Bajna - 0.432 and Jaora - 0.485) district is lower.

Field visits and interviews with the villagers have shown that the rate of degradation of natural resources was very high in the Malwa region. The natural resource degradation index is composed of three indicators and the analysis further revealed that time spent for collection of forest products and incidence of forest fire are increasing; and forest cover is desecrating. In Dhar district, households were solely dependent on the forest for fuel based solely. Most of the villages were surrounded by large chunks of open, barren land that was previously forested land and is now converted into agricultural land. Therefore, the natural resource degradation index is higher in Dhar (01.00) and Gandhwani (01.00) blocks. On the other side, vulnerability in the blocks of Ratlam district is lower (Jaora - 0.197 and Bajna - 0.452) due to comparatively lesser degradation and better management practices (Ann. 2,3,4 & 5). The higher rate of deforestation also degraded the water quality of natural sources. Along with this, higher runoff and lower percolation significantly affect the ground water level.

As discussed households of tribal blocks depend on forest for fodder and fuelwood. The nearby forest satisfies the requirement of fodder and fuelwood. Along with this, generally they allow livestock to graze in the outskirts of villages. This is one of the traditional practices and it reduces the pressure on fodder collection. On the other side, the requirement of fodder in non tribal blocks is fulfilled through using agricultural straw and the cultivation of fodder. The households of this block store some straw for emergency as a traditional practice, but the practice is completely lacking in tribal blocks. The better accessibility to fodder in non-tribal blocks (Dhar - 0.434 and Jaora - 0.433) in the strategy index showed lowers vulnerability.

Again, the strategies for accessibility to fuelwood during extreme events is better in non-tribal blocks (Dhar - 0.305 and Jaora - 0.308) because of the proximity to markets and access to better facilities. Most of the households of this block depend on Liqnefied Petroleum Gas (LPG) and kerosene. On the other side, villagers of tribal blocks solely depend on fuelwood and lack other alternatives like LPG and kerosene. Usually, they collect more fuelwood during the summer season and store it for the rainy season, but the increasing population pressure and decreasing forest cover directly affect sufficiency of fuelwood. The alternatives to fuelwood are lacking in tribal blocks such as Gandhwani (0.813) and Bajna (0.550) and hence they are more vulnerable.

Key Issues

1. Depletion of forest in order to expand agriculture
2. Unsustainable collection of fodder and fuelwood
3. Uncontrolled grazing
4. Spread of invasive species such as Parthenium and Lantana.

Existing/Ongoing Initiatives

1. Sustainable livelihood through Joint Forest Management (JFM): The degraded forest of the state is improved by involvement of village communities. They are entrusted with the protection, regeneration and management of nearby forests and they get rights for benefit sharing.

2. Management and protection of forests and afforestation on barren lands with the purpose of rural development.
3. Plantation on degraded and barren land by the forest department.
4. Compensation to villagers against loss of agriculture and cattle due to wild animals.

5.2 Vulnerability Assessment: IPCC Framework Approach

Vulnerability measures have been used extensively in interdisciplinary research to explain the degree to which a socio-economic and environmental system suffers from climate change. Therefore, it is necessary to appraise sector specific local vulnerability in order to create appropriate vulnerability reduction measures. The sector-specific or multiple-sector local vulnerability and adaptation measures in tribal as well as non tribal blocks of Dhar and Ratlam district of the Malwa region is already discussed in the first section. Further, it is equally important to assess the contribution of different dimensions (i.e., exposure, sensitivity and adaptive capacity) in the overall vulnerability of the Malwa region. Therefore, vulnerability of the Malwa region is also assessed using IPCC's vulnerability assessment approach, i.e., Vulnerability = (Exposure + Sensitivity) - Adaptive capacity.

The lower adaptive capacity of tribal households resulting from demographic imbalance, widespread poverty, and adaptation practices such as livelihood diversification and food scarcity and water storage, increased the overall vulnerability score. Although non tribal households reported comparatively better adaptation strategies, they did not report the school drop-outs and inaccessibility to facilities. The socio-economic status of non-tribal households was further strengthened by the possession of assets, better electricity and sanitation facilities, more irrigated land per households, and 12 months drinking water supply/facility inside the village boundary. The overall adaptive capacity of Dhar block, which represents the tribal block in the study, is 0.297; wherein Sukheda (0.267) village reported higher and Baggad (0.352) reported the lowest adaptive capacity (Annex 6). The adaptive capacity of the second tribal block, i.e., Jaora, is 0.244, and Rojana (0.18) and Patakhedi (0.287) villages reported the highest and lowest adaptive capacities. The adaptive capacity of the first tribal blocks, i.e., Gandhwani, is 0.523 and it ranges from 0.462 to 0.581. The second tribal block, Bajna, reported a slightly higher adaptive capacity, i.e., 0.471

than the first tribal block, wherein Mandaliya village reported the highest and Maanpur villages reported the lowest adaptive capacities.

Climate exposure indicators include biophysical factors such as temperature, heavy rain and drought. The IPCC predicts that the impact of global warming will continue as the probability of severe heat waves, heavy rain, drought, tropical depression, and sea level rise increases over time (Parry *et al.*, 2005). The exposure due to change in temperature and rainfall is approximately same in Dhar and Ratlam districts. Almost all of the respondents accepted that the climate is changing, in terms of temperature rise (summer and winter); variation in patterns, and frequency and intensity of rainfall. Climatic stress is, however, slightly higher in the tribal blocks, especially in Gandhwani, because of high exposure (0.631). The villages of Gandhwani block are in proximity to forested areas and they collect different types of resources such as fodder and fuelwood from the forest. But now the forest is degrading at an alarming rate and their dependency on the forest is increasing. The demand gap and pressure from the forest department cause psychological stress among the villagers. Additionally, extreme climatic events put more stresses due to change in the time/duration of resource collection. In contrast, a smaller proportion of respondents in non-tribal blocks were dependent on the forest for resources such as fodder and fuelwood and hence stress, as well as exposure, is lower in non tribal blocks (Dhar - 0.531 and Jaora – 0.412) (Annex 6).

The degree of a system's sensitivity to climatic hazards depends not only on geographic conditions but also socio-economic factors such as household dependency and poverty. The higher sensitivity of tribal blocks resulting from geographical conditions; water unavailability; demographic characteristics; natural resource degradation and lesser agricultural yield. The Gandhwani block reported the highest sensitivity, i.e., 0.584 and it is a result of the geographical location of the villages. Sensitivity of the block ranges from 0.424 to 0.717 and the highest sensitivity was reported by Gungi Devi village (0.424). The sensitivity of the second tribal block, i.e., Bajna, is slightly higher from that of the first tribal block. The developmental facilities in the block are constrained by varied geographical location; terrain and its parameters, that not only cut - off from the mainstream of development, but also cause hindrance in the villagers' accessibility. Further, the agriculture of the block is highly sensitive to irrigation facilities and extreme climatic events. On the other side, comparatively better geographical location, agricultural constraints and the improved standard of living by creating livelihood opportunities reduced the sensitivity of non tribal blocks. Among the non tribal blocks, Dhar block (0.219) reported slightly higher sensitivity and it ranges from 0.146 to 0.276. While the sensitivity of Jaora block is 0.169, the highest sensitivity was reported in Banbara village (0.224). The overall vulnerability of tribal blocks (Gandhwani – 0.693 and Bajna – 0.643) was higher as compared to non tribal blocks (Dhar - 0.453 and Jaora – 0.338). The overall higher vulnerability of tribal blocks might be attributed to higher exposure and more sensitivity (poverty; family dependency; migration and geographical location) with lower adaptive capacity (assets, electricity and sanitation, irrigation facility, drinking water supply/facility).

5.3. Prioritization of Components through Dimension Reduction Technique

The study uses the indicator based approach further grouped into ten components and 45 indices, which cover major aspects of rural livelihood. The data was collected on a number of variables and it is believed that there is some redundancy in the variables. This means some of the variables are correlated with one another, possibly because they are measuring the same construct. Therefore, to reduce the observed variables into a smaller number, principal component analysis (PCA) was used, that will account for most of the variance in the observed variables. Moreover, PCA would identify the most important factor (with the highest variance) that contributes significantly to sustainable livelihood. This may help to prioritize the components while planning for reducing the vulnerability.

Prior to the extraction of the factors, Kaiser-Meyer-Olkin (KMO) and Bartlett's Test of Sphericity were applied to assess the suitability of the survey data for factor analysis. The KMO metric, in particular, is recommended when the cases to variable ratio are less than 1:5. The KMO metric ranges from 0 to 1, with 0.50 considered suitable for factor analysis. The Bartlett's Test of Sphericity should be significant. The KMO measure of sampling adequacy for the present study is 0.728, above the recommended value of 0.5, and Bartlett's test of sphericity is significant ($\chi 2$ (45) = 304.49, $p < .05$). The eigenvalue of three factors was more than one. The eigenvalues showed that the first factor alone explained 49.16% of the variance, the second factor 16.10% of the variance, and a third factor 10.94% of the variance.

The eigenvalues of the remaining factors is less than one in each factor, explaining less than 10% of the variance. The three factor solutions, which explained 76.27% (Annex 11) of the variance, were preferred because of its previous theoretical support, the 'leveling off' of eigenvalues on the scree plot.

The first principal component is strongly correlated with four of the original variables and the component can be viewed as a measure of agriculture, socio-economic profile, social network, and education (Annex 12). The most influential factor is agriculture with the highest loadings, followed by education, socio-economic status and social network. All of the factors are related to the financial status of the household and could possibly be taken as *financial and social assets*. The strengthening of one factor probably enhances the status of others. As discussed, agriculture is the prime profession as well as source of livelihood for the villagers. The advancement of agriculture significantly improves the socio-economic profile of the villagers. A better socio-economic condition would break the financial constraints for education, and school as well as college enrollment could be increased. And the higher education and awareness may further promote good social networking among the villagers.

The second component also consists of four factors, i.e., food; water; natural resources; and weather perception and calamities. The highest and most influential factor is food with a loading of 0.478. The second most influential factor is natural resources with a factor loading of 0.455 followed by weather perception and calamities, and water. All the factors could collectively be *natural assets* and a change in one factor probably enhances the status of others (Annex 12).

The third component is composed of two variables, i.e., health and infrastructure. Health is the most influential factor in the third component with a factor loading of 0.674. All of the factors are related to a household's accessibility to basic facilities and the factor could possibly be taken as *physical assets*. The factors of this component are inter-linked with each other and strengthening of one factor significantly changes the state of the others. The infrastructural development at the village level comprises construction of government buildings and roads leading to better accessibility to facilities such as markets, banks, hospitals etc. Further, the better connectivity would also promote the regular and timely visits of medical staff, i.e., Asha worker, doctor etc.

References

Abramovitz J, Banuri T, Girot PO, Orlando B, Schneider N, Spanger-Siegfried E, Switzer J, Hammill A. (2001). Adapting to climate change: Natural resource management and vulnerability reduction. Gland: IUCN, Worldwatch Institute, IISD, Stockholm Environment Institute/ Boston (SEI-B)

Adger WN, Arnella NW, Tompkins EL. (2005). Successful adaptation to climate change across scales. Global Environmental Change 15 (2005) 77–86.

Adger WN, Brooks N, Bentham G, Agnew M, Eriksen S. (2004). *New indicators of vulnerability and adaptive capacity.* Technical report 7. Tyndall Centre for Climate Change Research.

Adger WN, Huq S, Brown K, Conway D, Hulme M. (2003) Adaptation to climate change in the developing world. Progress in Development Studies 3, 3 (2003) pp. 179–195.

Adger WN. (1996). Approaches to vulnerability to climate change. Centre for Social and Economic Research on the Global Environment Working Paper GEC 96–05. Norwich, U.K.: University of East Anglia.

Adger WN. (1999). Social Vulnerability to Climate Change and Extremes in Coastal Vietnam. World Development Vol. 27, No. 2, pp. 249-269.

Adger WN. (2000). Institutional adaptation to environmental risk under the transition in Vietnam. Annals of the Association of American Geographers 90, 738–758.

Adger WN. (2003). Social capital, collective action and adaptation to climate change. Economic Geography 79, 387–404.

Aggarwal PK. (2008). Global climate change and Indian agriculture: impacts, adaptation and mitigation, Indian Journal of Agricultural Sciences 78 (11): 911-919.

Bakkes JA, Van den Born GJ, Helder JC, Swart R J, Hope CW, and Parker JDE. (1994) An Overview of Environmental Indicators: State of the Art and Perspectives. UNEP/EATR.94-01; RIVM/402001001. Environmental Assessment Sub-Programme; UNEP, Nairobi

Blaikie P, Cannon T, Davis I, Wisner B. (1994). At Risk: Natural Hazards, People's Vulnerability, and Disasters. London: Routledge

Bohle HG. (2001). Vulnerability and Criticality: Perspectives from social geography. IHDP Update, Newsletter of the International Human Dimension Programme on Global Environmental Change (IHDP) 2:1–4

Brooks N. (2003). *Vulnerability, risk and adaptation: a conceptual framework.* Working Paper 39. Tyndall Centre for Climate Change Research.

Brooks N. (2003). Vulnerability, risk and adaptation: A conceptual framework. Working paper 38. Norwich, U.K.: Tydall Centre for Climate change Research, University of East Anglia

Cardona OD. (1990). Terminología de Uso Común en Manejo de Riesgos. AGID Reporte No. 13, Escuela de Administración, Finanzas, y Tecnología, Medellín, Colombia.

Census of India. (2011). Houselisting and Housing Census Schedule. Ministry of Home Affairs. Office of Registrar General and Census Commissioner, Government of India. Retrieved 23- 08-2013.

Cruz RV, H Harasawa, M Lal, Wu S, Anokhin Y, Punsalmaa B, Honda Y, Jafari M, Li C, Ninh NH. (2007). Climate Change 2007: Impacts, Adaptation and Vulnerability. Contribution of Working Group II to the Fourth Assessment Report of the Intergovernmental Panel on Climate Change [Parry, M.L., O.F. Canziani, J.P. Palutikof, P.J. van der Linden and C.E. Hanson (Eds.)], Intergovernmental Panel on Climate Change (IPCC), Cambridge: Cambridge University Press, 2007, p. 468

Das K. (2003-04). "Drinking Water and Sanitation in Rural Madhya Pradesh: A Review of Policy Initiatives". Ahmedabad: Gujarat Institute of Development Research.

Daze A, Ambrose K, Ehrhart C. (2009). *Climate vulnerability and capacity analysis handbook.* First edition. CARE International.

Deems H J, (2010). Vulnerability of rural communities in the Mediterranean region to climate change and water scarcity: The case of Cyprus. Dissertation, Universitat Autonoma de Barcelona, U.S.

Deressa T, Hassan RM, AlemuT, Yesuf M, Ringler C. (2008). Analysing the Determinants of Farmers' Choice of Adaptation Methods and Perceptions of Climate Change in the Nile Basin of Ethiopia, IFPRI Discussion Paper 00798: International Food Policy Research Institute.

Dolan AH, Walker IJ. (2003). Understanding Vulnerability of Coastal Communities to Climate Change Related Risks. Journal of Coastal Research SI 39: 0749-0208.

Dolan H, Walker I. (2004). Understanding vulnerability of coastal communities to climate change related risks. J. Coast. Res. Spec. Issue 39. Proceedings of the 8th International Coastal Symposium, Itajai, SC, Brazil.

Douglas EM, Niyogi D, Frolking S, Yeluripati JB, Pielke R A Sr, Niyogi N, Vörösmarty CJ, Mohanty UC. (2006). Changes in moisture and energy fluxes due to agricultural land use and irrigation in the Indian Monsoon Belt. Geophysics Research Letter;33:L14403. http://dx.doi.org/10.1029/2006GL026550.

Douglas EM, Beltrán-Przekurat A, Niyogi D, Pielke RA Sr, Vörösmarty CJ. (2009). The impact of agricultural intensification and irrigation on land–atmosphere interactions and Indian monsoon precipitation: a mesoscale modeling perspective. Global Planet Change;67:117–28.

Dow K. (1992). Exploring differences in our common future(s): the meaning of vulnerability to global environmental change. Geoforum. 23.3 p 417-436.

Ebert A, Kerle N, Stein A. (2008). Urban social vulnerability assessment with physical proxies and spatial metrics derived from air- and spaceborne imagery and GIS data. *Natural Hazards*, 48(2): 275–294.

Ebi K, Kovats RS, Menne B. (2006). An approach for assessing human health vulnerability and public health interventions to adapt to climate change. Environmental Health Perspectives 114(12): 1930-1934.

Engle NL. (2011). Adaptive capacity and its assessment. Global Environmental Change, 21: 647–656.

Eriksen SH, Kelly PM. (2006). Developing credible vulnerability indicators for climate adaptation policy assessment. *Mitigation and Adaptation Strategies for Global Change*, 12(4):495–524.

Esteves T, Rao KV, Sinha B, Roy SS, Rao B, Jha SK, Singh AB, Vishal P, Nitasha S, Rao S, Murthy IK, Sharma R, Porsche I, Basu K, Ravindranath NH. (2013). Agricultural and Livelihood Vulnerability Reduction through the MGNREGA. December 28, 2013. Vol XLVIII no 52. Economic & Political Weekly.

Fussel H M. (2007). Vulnerability: A Generally Applicable Conceptual Framework for Climate Change Research. Global Environmental Change 17(2): 155-167.

Fussel HM. (2007). Adaptation planning for climate change: concept, assessment approaches and key lessons. Integrated Research System for sustainability Science 2(2): 265-275.

Füssel HM. (2007). Vulnerability: a generally applicable conceptual framework for climate change research. Global Environmental Change, 17, 155-167

Füssel HM. (2007). Vulnerability: a generally applicable conceptual framework for climate change research. *Global Environmental Change*, 17(2): 155–167.

Gallopin GC. (1997). Indicators and their use: Information for decision-making. In *sustainability Indicators, Report of the Project on Indicators of Sustainable Development*; Moldan, B., Billharz, S., Eds.; John Wiley & Sons: Chichester, UK.

Gallopín GC. (2006). Linkages between vulnerability, resistance and adaptive capacity. Global Environmental Change 16(3): 293-303

Gallopín GC. (2006). Linkages between vulnerability, resistance and adaptive capacity. Global Environmental Change 16(3): 293-303

Gbetibouo GA, Hassan RM, Ringler C. (2010). Modelling Farmers' Adaptation Strategies for Climate Change and Variability: the Case of the Limpopo Basin, South Africa. Agrekon 49, pp. 217-234

Githeko AK, Woodward A. (2003). 'International consensus on the science of climate and health: The IPCC Third Assessment Report.' In McMicheal AJ, Campell-Lendrum DH, Corvalán CF, Ebi KL, Githeko AK, Scheraga JD, Woodward A. (eds) Climate change and human health – Risks and responses, pp 43–60. Geneva: World Health Organization

GIZ. (2014). The Vulnerability Sourcebook Concept and guidelines for standardized vulnerability assessments. BMZ Bonn, Germany.

Goulden M, Porter K. (2010). Updated CLICO conceptual framework. CLICO project deliverable D.1.1 report, Norwich: University of East Anglia, U.K.

Goulden M, Porter K. (2010). Updated CLICO conceptual framework. CLICO project deliverable D.1.1 report, Norwich: University of East Anglia, U.K.

Gregory PJ, Ingram JSI, Brklacich M. (2005). 'Climate change and food security.' Philosophical Transactions of the Royal Society of London. B 360: 2139–2148

Hahn MB, Riederer AM, Foster SO. (2009). The Livelihood Vulnerability Index: A pragmatic approach to assessing risks from climate variability and change—A case study in Mozambique. *Global Environmental Change*, 19(1): 74–88.

Hammond A, Adriaanse A, Rodenburg E, Bryant D, Woodward R. (1995). Environmental Indicators: A systematic approach to measuring and reporting on environmental policy performance in the context of sustainable development., World Resources Institute, Washington, D.C

Handmer JW, Dovers S, Downing TE. (1999). 'Societal Vulnerability to Climate Change and Variability' *Mitigation and Adaptation Strategies for Global Change* 4(3–4), 267–281.

Hanjra MA, Qureshi ME. (2010). 'Global water crisis and future food security in an era of climate change.' Food Policy 35: 365–377

Haughton J, Khandker SR. (2009). Handbook on poverty and inequality. Washington, DC: World Bank.

Hedger M, Vaideeswaran S. (2010). Scoping Study for Madhya Pradesh State Climate Change Action Plan. A report for the Government of Madhya Pradesh

Hinkel J. (2011). "Indicators of vulnerability and adaptive capacity": towards a clarification of the science–policy interface. *Global Environmental Change*, 21(1): 198–208.

Holling CS. (1978). Adaptive Environmental Assessment and Management. John Wiley & Sons,Chichester.

Human Rights Education Associates. (2012). The right to housing – Study guide. Amsterdam: HREA. http://www.hrea.org/index.php?doc_id=411.

Hunzai K, Gerlitz JY, Hoermann B. (2011). Understanding mountain poverty in the Hindu Kush-Himalayas: Regional report for Afghanistan, Bangladesh, Bhutan, China, India, Myanmar, Nepal, and Pakistan. Kathmandu: ICIMOD.

IHDR. (2011). India Human Development Report 2011: Towards Social Inclusion. Institute of Applied Manpower Research, Planning Commission, Government of India.

IIPS. (2010). District Level Household and Facility Survey (DLHS-3), 2007-08, Madhya Pradesh. International Institute for Population Sciences (Deemed University) Mumbai-400088.

INCCA. (2010). Climate change and india: a 4X4 assessment a sectoral and regional analysis for 2030s. Government of India.

IPCC. (2000). Report by Robert Watson, Chair, Intergovernmental Panel on Climate Change, to the Sixth Conference of the Parties to the United Nations Framework Convention on Climate Change, The Hague, 20 November 2000.

IPCC. (2001). Climate Change 2001, Impacts, Adaptation and Vulnerability. Summary for Policymakers

IPCC. (2001). Climate Change 2001, Impacts, Adaptation and Vulnerability. Summary for Policymakers

IPCC. (2001). Climate Change 2001: Impacts, Adaptation and Vulnerability. Contribution of Working Group II to the Third Assessment Report of the Intergovernmental Panel on Climate Change, pp. 981–996. J.J. McCarthy, O.F. Canziani, N.A. Leary, D.J. Dokken & K.S. White, eds. Cambridge, UK and New York, USA, Cambridge University Press.

IPCC. (2001). Climate Change 2001: Impacts, Adaptation Vulnerability. Contribution of Working Group II to the Third Assessment Report of the Intergovernmental Panel on Climate Change. Geneva: UNEP/WMO.

IPCC. (2014). Summary for policymakers. In: Climate Change 2014: Impacts,Adaptation, and Vulnerability. Part A: Global and Sectoral Aspects. Contribution of Working Group II to the Fifth Assessment Report of the Intergovernmental Panel on Climate Change [Field, C.B., V.R. Barros, D.J. Dokken, K.J. Mach, M.D. Mastrandrea, T.E. Bilir, M. Chatterjee, K.L. Ebi, Y.O. Estrada, R.C. Genova, B. Girma, E.S. Kissel, A.N. Levy, S. MacCracken, P.R. Mastrandrea, and L.L.White (eds.)]. Cambridge University Press, Cambridge, United Kingdom and New York, NY, USA, pp. 1-32.

Kates RW, Ausubel JH, Berberian M. (eds.) (1985). Scope Report on Impact Assessments. Wiley: UK.

Keim M. (2008). Building Human Resilience: The Role of Public Health Preparedness and Response As an Adaptation to Climate Change. American Journal of Preventive Medicine , 5 (35), 508-516.

Kelkara U, Narulab KK, Sharma VP, Chandna U. (2008).Vulnerability and adaptation to climate variability and water stress in Uttarakhand State, India. Global Environmental Change Volume 18, Issue 4, October 2008, Pages 564–574.

Kelly PM, Adger WN. (2000). Theory and practice in assessing vulnerability to climate change and facilitating adaptation. Climatic Change 47: 325–352.

Kothawale DR, Munot AA, Kumar KK. (2010). Surface air temperature variability over India during 1901-2007, and its association with ENSO, Climate Research, 42, June 2010, 89- 104, doi:10.3354/ cr00857.

Kumar KSK, Parikh J. (2001). Indian Agriculture and Climate Sensitivity. Global Environmental Change. 11: 147-154.

Marcus S. (1983). Towards a Semiotic Approach to Social Indicators. Paper presented at the International Congress of Logic, Methodology and Philosophy of Science, Salzburg, Austria, July 11-16,1983.

Maxwell S, Smith M. (1992). 'Household food security: A conceptual review.' In Maxwell S, Frankenberger T. (eds) Household Food Security: Concepts, Indicators, and Measurements, pp 1–71. Rome: UNICEF and IFAD.

Maxwell S, Smith M. (1992). Household food security; a conceptual review. In S. Maxwell & T.R. Frankenberger, eds. Household Food Security: Concepts, Indicators, Measurements: A Technical Review. New York and Rome: UNICEF and IFAD.

McCarthy J J, Canziani O F, Leary N A, Dokken D J, White K S (ed). (2001). Climate change 2001: Impacts, adaptation and vulnerability. Cambridge, UK: Cambridge University Press

McKenzie DJ. (2005). 'Measuring inequality with asset indicators.' Journal of Population Economics 18: 229–260

McMichael A, Woodruff R. (2005). Climate change and human health. Netherlands: Springer.

McQueen D, Noak H. (1988). Health Promotion Indicators: Current status, issues and problems. Health Promotion 3, 117-125. (MWCD)

McSweeney K, Coomes O T. (2011). Climate-related disaster opens a window of opportunity for rural poor in northern Honduras. PNAS 108(13): 5203-5208

Mertz O, Mbow C, Reenberg A. (2008). Farmers' Perception of Climate Change and Agricultural Adaptation Strategies in Rural Sahel. Environmental Management, 43: 804- 816.

Molden D. (2007). 'Water responses to urbanization.' Paddy and Water Environment 5: 207–209.

Moss RH, Brenkert AL, Malone EL. (2001). Vulnerability to Climate Change: A Quantitative Approach. Dept. of Energy, US.

MPDR. (2011). Madhya Pradesh Development Report. Planning Commission, Government of India, New Delhi. India

MP-SAPCC. (2013). Madhya Pradesh State Action Plan on Climate Change. Housing and Environment Department. Government of Madhya Pradesh.

Munang R, Rivington M, Takle ES, Mackey B, Thiaw I, Liu J. (2010). Climate Information and Capacity Needs for Ecosystem Management under a Changing Climate. Procedia Environmental Sciences 1(2010): 206–227

O'Brien KL, Eriksen S, Schjolden A, Nygaard L. (2004). What's in a word? Conflicting interpretations of vulnerability in climate change research. CICERO Working Paper 2004:04. Oslo, Norway: Centre for International Climate and Environmental Research.

OECD. (1993). Organization for Economic Cooperation and Development Core Set of Indicators for Environmental Performance Reviews. A Synthesis Report by the Group on the State of the Environment. OECD, Paris.

Ott WR. (1978). Environmental Indices: Theory and Practice. Ann Arbor Science, Michigan.

Ozor N, Madukwe MC, Enete AA, Amaechina EC, Onokala P, Eboh E C, Ujah O, Garforth, C. (2012). *A framework for agricultural adaptation to climate change in Southern Nigeria.* International Journal of Agriculture Sciences, 4 (5). pp. 243-251. ISSN 0975-3710

Pandey R, and Jha SK. (2012) Climate vulnerability index - measure of climate change vulnerability to communities: a case of rural Lower Himalaya, India. Mitigation and adaptation and strategies for global change. Springer, Volume 17, Issue 5, pp 487-506.

Paroda RS, Kumar P. (2000). Food production and demand in South Asia, Agricultural Economics Research Review 13(1): 1-24.

Pearson LJ, Nelsonc R, Crimp S. Langridge J. (2011). Interpretive review of conceptual frameworks and research models that inform Australia's agricultural vulnerability to climate change. *Environmental Modelling & Software*, 26(2): 113–123.

Ragab R, Prudhomme C. (2002). 'Climate change and water resources management in arid and semi-arid regions: Prospective and challenges for the 21st Century.' Biosystems Engineering 81(1): 3–34

Rodell M, Velicogna I, Famiglietti JS. (2009). Satellite-based estimates of groundwater depletion in India, Vol 460. ; 2009. http://dx.doi.org/10.1038/nature08238.

Samra J S, Singh G. (2004). Heat wave of March 2004: Impact on agriculture, Indian Council of Agricultural Research. 32 pp.

Schmidhuber J, Tubiello FN. (2007). 'Global food security under climate change.' Proceedings of the National Academy of Sciences 104 (50): 19703–19708.

Selvaraj R. (1998). "Commercial Bank Finance: Rural Customer's Satisfaction", Kurushetra, October 1998, pp. 44 – 45.

Sen AK. (1981). *Poverty and famines: An essay on entitlement and deprivation.* Clarendon, Oxford.

SFP. (2014-15). State Focus Paper, Madhya Pradesh. NABARD-RO, Bhopal, Madhya Pradesh. India.

Singh PK, Nair A. (2013). Climate Change Vulnerability Assessment for Sustainable Livelihoods using Fuzzy Cognitive Mapping Approach. Working Paper 248. Institute of Rural Management, Anand,Gujarat (India)

Sinha B, Jha SK, Khare AK. (2015). Assessing the potential of incentive based plantation involving people's participation. Indian Forester, 141 (4): 439-447, 2015.

Smit B, Wandel J. (2006). Adaptation, adaptive capacity and vulnerability. Global Environmental Change 16(3): 282:292

Smithers J, Barrett C B. (1997). Human adaptation to climatic variability and change. Global Environmental change 7(2): 129-146

Sullivan CA. (2002). 'Calculating a water poverty index', *World Development*, vol. 30, no. 7, pp. 1195-1210.

Thomas DSG, Twyman C, Osbahr H, Hewitson B. (2007). Adaptation to climate change and variability: farmer responses to intra-seasonal precipitation trends in South Africa. Climate Change. August 2007, Volume 83, Issue 3, pp 301-322

Tol R, Ebi KL, Yohe G. (2007). Infectious disease, development, and climate change: a scenario analysis. Environment and Development Economics, 2007, vol. 12, issue 05, pages 687-706.

Tunstall D. (1992). Developing Environmental Indicators: Definitions, framework and issues. (Draft paper). Background Materials for the World Resources Institute. Workshop on Global Environmental Indicators, Washington, D.C., December 7-8, 1992. World Resources Institute, Washington, D.C.

UNISDR. (2009). Terminology on Disaster Risk Reduction. United Nations International Strategy for Disaster Reduction, Geneva, Switzerland. unisdr. org/eng/library/lib-terminology-cng.htm.

Urothody AA, Larsen HO. (2010). Measuring climate change vulnerability: a comparison of two indexes. Banko Jankari, Katmandu Nepal, 20 (1) 9-16.

Vincent K, Cull T. (2010). A Household Social Vulnerability Index (HSVI) for Evaluating Adaptation Projects in Developing Countries. PEGNet Conference 2010: Policies to foster and sustain equitable development in times of crises, Midrand, 2-3rd September 2010

Vincent K. (2004). Creating an Index of Social Vulnerability to Climate Change for Africa Tyndall, Centre for Climate Change Research Working Paper 56.

Vincent K. (2007). Uncertainty in adaptive capacity and the importance of scale. Global Environmental Change 17 (2007) 12–24.

Walker B, Holling CS, Carpenter SR, Kinzig A. (2004). Resilience, Adaptability and Transformability in Social-ecological Syatems. Ecology and Society 9 (2):5.

Watson RT, Zinyoera MC, Moss RH. (1996). Climate Change 1995: Impacts, Adaptations and Mitigation of Climate Change: Scientific-Technical Analysis. Contribution of Working Group II to the Second Assessment Report of the Intergovernmental Panel on Climate Change. Cambridge: Cambridge University Press.

WHO. (2006). Constitution of the World Health Organization. Basic Documents (Forty-fifth edition, Supplement). Geneva: WHO

Xu J, Sharma R, Fang J, Xu Y. (2008). Critical linkages between land-use transition and human health in the Himalayan region. Environment International, 34 (2008) 239-247.

Website:

http://www.freepressjournal.in/quantum-jump-in-gdp-per-capita-income-of-state-in-2014-15/ (02-09-2015)

http://www.dailypioneer.com/state-editions/bhopal/gsdp-growth-mp-likely-to-outdo-other states. html(02-09-2015)

Appendix

Annex. 1: Explanation of Components of Vulnerability and Indicators

Components	Indexes	Indicators	Explanation	References
Socio-Economic Profile	Family dependency index	Number of dependent (upto 18 and above 55) per household.	Higher dependency in a household depicts more mouth to eat and less hands to work and hence more vulnerability.	Hahn *et al.*, 2009; Pandey *et al.*, 2011; Joerin *et al.*, 2014; Gerlitz *et al.*, 2014; Madhury *et al.*, 2014; Sinha *et al.*, 2015,
	Profession diversity index	The inverse of (types of profession of household +1) reported by a household.	A household with different source of livelihood might have better livelihood strategy.	Pandey *et al.*, 2011
	Poverty line index	Percentage of household below poverty line.	Poor & marginalized are more vulnerable.	Cruz *et al.*, 2007; Stern 2007.
	House type diversity index	The inverse of (types of house +1) reported by a household.	Households with non concrete house may have less adaptive capacity.	Pandey *et al.*, 2011; Sinha *et al.*, 2015
	Asset index	Percentage of household with basic assets.	Households with basic assets such as bicycle, Motorbike, tractor, refrigerator, TV, radio, etc. is less susceptible.	Gerlitz *et al.*, 2014
	Livestock index	Number and type of livestock per household.	Household with different types and more number of livestock reflects better adaptive capacity.	Hahn *et al.*, 2009; Pandey *et al.*, 2011; Esteves, 2013

	Migration index	Percentage of household with migrated members.	Migration indicates sensitivity of household.	Gerlitz *et al.*, 2014; Esteves, 2013
		Number of member migrated (at least 1)	Household with more migration indicates higher sensitivity.	Gerlitz *et al.*, 2014; Esteves, 2013
	Electricity index	Percentage of household having electricity connection	Access to electricity has implications for the health, education, and income of households.	Joerin *et al.*, 2014
		Availability (hr/day) of electricity.	Accessibility along with availability (hr/day) of electricity is important for livelihood activities.	Joerin *et al.*, 2014
	Sanitation index	Percentage of household without toilet facility.	Household having sanitation facility are less vulnerable to diseases.	Madhury *et al.*, 2014; Joerin *et al.*, 2014; Gerlitz *et al.*, 2014
	Stability index	Percentage of household having policy or insurance.	Policy or insurance indicates saving as well as awareness of a household.	
Social Network	Assistance index	Percentage of household reported to have received help from relatives and friends	Households get assistance during extreme condition is less vulnerable.	Pandey *et al.*, 2011
		Percentage of household does not provide assistance to other during extreme condition.	Household does not help others indicates poor social network.	Pandey *et al.*, 2011
	Money transaction index	Money transaction between communities	Money transaction between communities reflects social cohesion.	

		Source of loan for a household (lender, bank, SHG, neighbor, relative)	Vulnerability might be higher, if they takes loan from lender or rate of interest is high.	
		Purpose of loan for a household (medical, education, marriage /occasion and agriculture)	Vulnerability might be higher, if a household takes loan for medical and marriage/ occasion.	
Education	Youth education index	Percentage of un-educated youth (below 18).	Reduced access to lifelong learning and professional development.	Joerin *et al.*, 2014
	Household education index	Percentage of uneducated population.	Limited ability to obtain livelihood.	
	Household decision index	Percentage of household where head of household is female	Male can take better decision as compare to female.	Pandey *et al.*, 2011
		Percentage of household where head is illiterate	Literate male can take better decision as compared to illiterate male or literate/illiterate female.	Pandey *et al.*, 2011
		Percentage of household where head of household is less than 40 years of age.	Head of the household with more than 40 years of age can take better decision.	Pandey *et al.*, 2011
Agriculture	Agricultural equipments index	Percentage of household with agricultural equipments like tractor, power tiller, motor pump etc.	Basic agricultural equipments are of critical importance for agriculture and it reflects household's capacity.	

	Agricultural land index	Percentage of irrigated land	Households can take two or more corps in a year, if they posses irrigated land.	Esteves, 2013
	Cropping intensity index	Cropping intensity of a household	Higher the cropping intensity greater the production.	Esteves, 2013
	Irrigation index	Percentage of household having irrigation facility	Household's own source of irrigation indicates that they are not dependent on others for irrigation.	
		Source of irrigation	Perennial source of irrigation is comparatively better.	
		Availability (Months) of irrigation	The availability of water in the sources is more leads to more water for irrigation.	Esteves, 2013
		Number of irrigation	Yield of rabi crop may also be depends on number of irrigation.	Esteves, 2013
	Irrigation strategy index	Strategies during shortage of irrigation (plant less area, request for government support and change crop).	Households with more strategies for irrigation reflect lesser susceptibility.	
	Agricultural strategy index	Percentage of household reported forward/backward sowing/planting time	Household's awareness and assess to weather forecast reflected by change in time for sowing/harvesting.	
		Percentage of household changes irrigation pattern	Households changes irrigation pattern as per water availability are more aware and less susceptible.	

		Percentage of household changes crop variety.	Households changes crop variety in accordance with weather prediction are more aware and less susceptible.	
		Percentage of household changes in crop type	Households changes type of crop in accordance with weather prediction are more aware and less susceptible	
Food	Food sufficiency index	Primary source (PDS, agriculture, others) of food	Households with more than one source of food reflect higher adaptive capacity.	
		Sufficiency of food across the year.	Households get sufficient food across the year is less vulnerable.	Pandey *et al.*, 2011; Sinha *et al.*, 2015
	Nutritious food index	Variation in consumption of nutritious food in last 10 years.	Households with lesser availability of nutritious food items like milk, fruit, legumes, meat and fish are more vulnerability.	Pandey *et al.*, 2011
Water	Portable water Accessibility Index	Sources of portable water	Perennial source of water may satisfy the water requirement for longer period.	Joerin *et al.*, 2014
		Source of portable water during lean season or alternative source of drinking water	Perennial or alternate source of water during lean season reflects better water accessibility.	Joerin *et al.*, 2014

	Portable water availability index	Availability of portable water (Months) inside the village or out of village	Water availability inside the village boundary reduces time and distance for water collection.	Joerin *et al.,* 2014
		Availability of portable water (Months) during lean season in the village or out of village	Water availability inside the village boundary reduces time and distance for water collection during lean season.	Joerin *et al.,* 2014
	Portable water collection index	Time spent on water collection	The source of water inside the village boundary leads to lesser time spent on water collection.	Hahn *et al.,* 2009; Gerlitz *et al.,* 2014
		Time spent on water collection during lean season	More time spent on water collection, if the alternate source of water away from village.	
		Distance travelled for water collection	More distance travelled for water collection, if the source of water away from village.	
		Distance travelled for water collection during lean season	The alternate source of water away from village leads to more distance travelled for water collection.	
Weather Perception and Calamities	Weather Forecast index	Access to weather forecast	Better access to weather forecast higher the chance of reliability.	
		Reliability of weather forecast	Higher the reliability of forecast lesser the chance of loss.	

	Summer temperature perception index	Variation in summer temperature	More variation in temperature may leads to greater loss.	Pandey *et al.*, 2011
		Frequency of hot months	Increase in frequency of hot month's leads to leads to greater loss.	Pandey *et al.*, 2011
	Rainfall perception index	Change in frequency of rainfall	Change in frequency of rainfall leads to greater loss.	Pandey *et al.*, 2011
		Change in intensity of rainfall	Change in intensity of rainfall leads to greater loss.	Pandey *et al.*, 2011
		Change in rainfall pattern	Change in pattern of rainfall leads to greater loss.	Pandey *et al.*, 2011
	Winter temperature perception index	Change in winter temperature	Change in winter temperature leads to more prolonged summer, it may affect cropping pattern.	Pandey *et al.*, 2011
	Natural Calamities index	Increase in frequency of drought	More frequent drought significantly affect livelihood especially agriculture.	
		Increase in frequency of hailstrom	More frequent hailstrom significantly affect livelihood especially agriculture.	Pandey *et al.*, 2011
		Increase in frequency of cold waves.	More frequent cold waves significantly affect livelihood especially agriculture.	Pandey *et al.*, 2011
	Stress Index	Stress due to rainfall	Stress may affect the efficiency of villagers.	
		Stress due to temperature	Stress may affect the efficiency of villagers.	

	Climatic disease index	Disease due to climatic variation	Climate diseases might be the additional burden for household.	
		Any new disease	New diseases in last 05 to 10 years may affect the resilience of household.	Pandey *et al.*, 2011
Natural Resources	Natural resource availability index	Percentage of household collect fodder and fuel-wood from forest	Households collect fodder and fuel-wood from forest shows their dependency on forest. Although forest cover in district is very low.	
		Percentage of household does not get sufficient fodder and fuel-wood.	Forest is deteriorating at alarming rate.	Sinha *et al.*, 2015.
	Natural resource collection index	Distance travelled for collection of fodder and fuel-wood	The distance between forest and village is increasing depicts that forest is deteriorating at alarming rate.	
		Time spent on collection of fodder and fuel-wood	They have to spent additional time for on collection.	
	Natural resource degradation index	Household reported time spent on fodder and fuelwood collection is increasing	The average time for firewood collection has been increased represents either household's dependency on forest is increasing or forest cover is being decreasing.	
		Household reported forest cover is decreasing	Higher rate of forest degradation.	
		Household reported more frequent forest fire	Higher incidence of forest fire reflect higher rate of degradation.	

	Natural water quality index	Change in quality of natural water.	Higher rate of forest degradation leads to lesser ground water recharge.	
	Fodder strategy index	Strategies during shortage of fodder (cultivation of fodder or use of more straw)	Household with more strategy for fodder are less susceptible.	
	Fuelwood strategy index	Strategies during shortage of fuelwood (shifting to other fuel like kerosine, coal and LPG)	Household with more alternatives for fuel wood are less susceptible.	
Health	Health worker index	Frequency of Health worker (Asha) visit/week in the village	More frequent visit of health worker leads to more health related awareness among villagers.	
		Frequency of Doctor's visit/week in the village	Regular check-up and better access to medical facility significantly affect the adaptive capacity of household.	
	Health facility accessibility index	Distance from medicine shop	More distance less inclination.	
		Distance from community health center	Time to reach hospital has been increased leads to lower inclination towards proper healthcare and higher dependency on local vaidya.	Hahn *et al.*, 2009; Joerin *et al.*, 2014; Gerlitz *et al.*, 2014
Infra-structure	Connec-tivity index	Road connectivity	Better connectivity leads to more frequent visit to market, community health center, bank etc.	

		Distance from bus stop	The bus stop is far from village leads to more time spent to reach bus stop and travelling.	
	Market accessibility index	Distance from nearest market	More time is required to reach market lead to be less inclination of villagers to go to market for selling or buying goods.	
	Bank accessibility index	Distance from bank branch	Lesser distance leads to easy and frequent accessibility to bank.	
	Govern-ment resource index	Villages possess primary school	Better access to education.	
		Villages possess Panchayat Bhawan	Panchayat Bhawan reflects the household's accessibility to institution.	

Annex 2: Indexed components and sub-components for vulnerability and capacity assessment of Dhar block

Components	Sub-Components	Antrai	Saktali	Lohari Bujurg	Baggad	Samar	Masawda	Sukheda	Nekpur	Sirsoda	Tornod	Dhar	Dh_VCA
Socio-Economic Profile	Family dependency index	0.520	0.344	0.250	0.380	0.390	0.440	0.630	0.330	0.280	0.415	0.398	0.282
	Profession diversity index	0.250	0.250	0.200	0.250	0.250	0.200	0.200	0.200	0.330	0.250	0.238	
	Poverty line index	0.500	0.330	0.200	0.300	0.400	0.500	0.375	0.100	0.000	0.100	0.281	
	House type diversity index	0.25	0.383	0.366	0.391	0.333	0.291	0.25	0.283	0.375	0.258	0.318	
	Asset index	0.143	0.205	0.307	0.410	0.407	0.152	0.126	0.158	0.188	0.125	0.222	
	Livestock index	0.369	0.333	0.263	0.310	0.276	0.250	0.232	0.514	0.235	0.213	0.300	
	Migration index	0.000	0.000	0.000	0.000	0.000	0.000	0.000	0.100	0.000	0.300	0.040	
	Electricity index	0.250	0.667	0.375	0.666	0.375	0.500	0.416	0.333	0.333	0.416	0.433	
	Sanitation index	0.000	0.266	0.300	0.400	0.200	0.300	0.000	0.300	0.300	0.000	0.207	
	Stability index	0.344	0.396	0.500	0.400	0.416	0.370	0.333	0.400	0.370	0.357	0.389	
Social Network	Assistance index	0.211	0.174	0.227	0.185	0.208	0.400	0.229	0.172	0.172	0.192	0.217	0.280
	Money transaction index	0.392	0.441	0.296	0.461	0.344	0.281	0.236	0.344	0.302	0.337	0.343	
Education	Youth education index	0.046	0.000	0.000	0.000	0.000	0.250	0.154	0.000	0.077	0.158	0.069	0.145
	Household education index	0.168	0.246	0.045	0.108	0.090	0.503	0.279	0.183	0.188	0.275	0.209	
	Household decision index	0.366	0.178	0.033	0.133	0.167	0.133	0.133	0.100	0.200	0.133	0.158	
Agriculture	Agricultural equipments index	0.181	0.111	0.550	0.400	0.220	0.235	0.228	0.125	0.129	0.137	0.232	0.203
	Agricultural land index	0.058	0.120	0.370	0.040	0.131	0.054	0.015	0.031	0.017	0.100	0.094	
	Cropping intensity index	0.140	0.125	0.593	0.200	0.389	0.200	0.125	0.125	0.000	0.100	0.200	

	Irrigation index	0.265	0.286	0.235	0.286	0.272	0.236	0.237	0.219	0.234	0.247	0.252	
	Irrigation strategy index	0.256	0.268	0.347	0.368	0.310	0.384	0.363	0.360	0.370	0.370	0.340	
	Agricultural strategy index	0.091	0.102	0.127	0.100	0.101	0.094	0.142	0.091	0.091	0.091	0.103	
Food	Food sufficiency index	0.225	0.275	0.256	0.247	0.264	0.225	0.225	0.235	0.225	0.229	0.241	0.184
	Nutritious food index	0.060	0.102	0.073	0.094	0.066	0.062	0.076	0.615	0.060	0.073	0.128	
Water	Portable water Accessibility Index	0.167	0.464	0.417	0.375	0.450	0.135	0.108	0.083	0.375	0.320	0.289	0.238
	Portable water availability index	0.000	0.157	0.208	0.161	0.207	0.000	0.000	0.000	0.571	0.114	0.142	
	Portable water collection index	0.204	0.430	0.345	0.464	0.375	0.344	0.328	0.000	0.332	0.295	0.312	
Weather Perception and Calamities	Weather Forecast index	0.275	0.453	0.328	0.417	0.290	0.292	0.327	0.283	0.291	0.328	0.328	0.432
	Summer temperature perception index	0.500	0.456	0.477	0.467	0.477	0.455	0.500	0.450	0.488	0.500	0.477	
	Rainfall perception index	0.500	0.345	0.343	0.337	0.401	0.360	0.433	0.389	0.500	0.500	0.411	
	Winter temperature perception index	0.500	0.483	0.370	0.434	0.400	0.384	0.444	0.421	0.500	0.500	0.444	
	Natural Calamities index	0.500	0.440	0.436	0.434	0.485	0.387	0.500	0.455	0.500	0.500	0.464	
	Stress Index	0.500	0.500	0.500	0.500	0.500	0.400	0.500	0.450	0.500	0.476	0.483	
	Climatic disease index	0.488	0.433	0.417	0.435	0.442	0.359	0.389	0.357	0.411	0.443	0.417	
Natural Resources	Natural resource availability index	0.500	0.500	0.500	0.500	0.500	0.500	0.500	0.500	0.500	0.500	0.500	0.613
	Natural resource collection index	1.000	1.000	1.000	1.000	1.000	1.000	1.000	1.000	1.000	1.000	1.000	

	Natural resource degradation index	1.000	1.000	1.000	1.000	1.000	1.000	1.000	1.000	1.000	1.000	1.000	
	Natural water quality index	0.500	0.384	0.500	0.384	0.500	0.455	0.500	0.370	0.455	0.344	0.439	
	Fodder strategy index	0.344	0.416	0.455	0.409	0.416	0.500	0.500	0.434	0.370	0.500	0.434	
	Fuelwood strategy index	0.256	0.250	0.500	0.294	0.357	0.250	0.333	0.312	0.250	0.250	0.305	
Health	Health worker index	0.200	0.267	0.225	0.267	0.225	0.225	0.225	0.225	0.267	0.267	0.239	0.355
	Health facility accessibility index	0.733	0.267	1.000	0.817	0.233	0.550	0.467	0.400	0.233	0.000	0.470	
Infra-structure	Connectivity index	0.167	0.167	0.417	0.417	0.333	0.333	0.417	0.417	0.417	0.417	0.350	0.209
	Market accessibility index	0.000	0.000	0.000	0.000	0.000	0.000	0.000	0.000	0.000	0.000	0.000	
	Bank accessibility index	0.091	0.200	0.077	0.091	0.167	0.111	0.111	0.167	0.167	0.333	0.151	
	Government resource index	0.333	0.333	0.333	0.333	0.333	0.333	0.333	0.333	0.333	0.333	0.333	

Annex 3: Indexed components and sub-components for vulnerability and capacity assessment of Gandhwani block

Components	Sub-Components	Baleri	Gungi devi	Jalokya	Karcha	Ke-likhurd	Lo-harpura	Maalgarh	Machalli	Motimali	Panch-pipaliya	Gandh-wani	Gw_VCA
Socio-Economic Profile	Family dependency index	0.317	0.235	0.465	0.567	0.507	0.493	0.576	0.574	0.552	0.575	0.486	0.648
	Profession diversity index	0.333	0.500	0.333	0.333	0.500	0.333	0.333	0.500	0.333	0.333	0.383	
	Poverty line index	1.000	1.000	1.000	1.000	1.000	1.000	1.000	1.000	1.000	1.000	1.000	
	House type diversity index	0.458	0.500	0.325	0.500	0.466	0.455	0.483	0.500	0.500	0.483	0.467	
	Asset index	0.875	0.385	0.444	0.471	0.714	0.745	0.740	0.567	0.667	0.620	0.623	
	Livestock index	0.255	0.242	0.148	0.179	0.182	0.255	0.231	0.186	0.156	0.167	0.200	
	Migration index	0.762	0.534	0.535	0.717	0.825	0.725	0.663	0.459	0.710	0.609	0.654	
	Electricity index	0.666	0.666	0.750	0.750	0.583	0.750	0.750	1.000	0.750	0.750	0.742	
	Sanitation index	0.777	1.000	1.000	0.900	1.000	0.818	0.900	1.000	0.900	1.000	0.930	
	Stability index	1.000	1.000	1.000	1.000	1.000	1.000	1.000	1.000	1.000	1.000	1.000	
Social Net-work	Assistance index	0.250	0.550	1.000	0.905	0.208	0.212	0.905	1.000	0.908	0.583	0.652	0.605
	Money transaction index	0.632	0.557	0.467	0.564	0.465	0.505	0.548	0.767	0.502	0.575	0.558	
Education	Youth education index	0.111	0.400	0.526	0.585	0.424	0.371	0.666	0.500	0.529	0.500	0.461	0.511
	Household education index	0.649	0.779	0.793	0.727	0.539	0.504	0.711	0.754	0.746	0.562	0.676	
	Household decision index	0.333	0.266	0.466	0.400	0.500	0.454	0.400	0.400	0.300	0.433	0.395	
Agriculture	Agricultural equipments index	1.000	0.600	0.633	0.900	0.667	0.857	0.900	0.100	1.000	0.900	0.756	0.562
	Agricultural land index	0.723	0.456	0.790	0.897	0.403	0.941	0.862	0.925	1.000	0.764	0.776	
	Cropping intensity index	0.447	0.457	0.790	0.693	0.606	0.825	0.794	0.775	1.000	0.706	0.709	
	Irrigation index	0.32	0.279	0.464	0.509	0.274	0.667	0.581	0.644	1.000	0.491	0.523	
	Irrigation strategy index	0.421	0.500	0.500	0.526	0.500	0.478	0.526	0.500	0.500	0.500	0.495	
	Agricultural strategy index	0.129	0.125	0.123	0.120	0.125	0.011	0.120	0.125	0.120	0.120	0.112	
Food	Food sufficiency index	0.307	0.292	0.292	0.263	0.242	0.243	0.257	0.292	0.257	0.257	0.270	0.179
	Nutritious food index	0.070	0.121	0.088	0.088	0.086	0.090	0.088	0.091	0.090	0.078	0.089	

Water	Portable water Accessibility Index	0.500	0.167	0.500	0.500	0.417	0.433	0.250	0.250	0.250	0.417	0.368	0.409
	Portable water availability index	0.178	0.038	0.306	0.156	0.168	0.197	0.038	0.038	0.038	0.156	0.131	
	Portable water collection index	0.613	0.674	0.806	0.667	0.811	0.703	0.884	0.900	0.599	0.631	0.729	
Weather Perception and Calamities	Weather Forecast index	0.402	0.220	0.402	0.352	0.429	0.406	0.405	0.402	0.379	0.381	0.377	0.456
	Summer temperature perception index	0.500	0.500	0.500	0.488	0.500	0.489	0.488	0.500	0.500	0.500	0.497	
	Rainfall perception index	0.500	0.500	0.500	0.431	0.450	0.436	0.431	0.500	0.468	0.485	0.470	
	Winter temperature perception index	0.500	0.500	0.500	0.416	0.500	0.407	0.416	0.500	0.466	0.455	0.466	
	Natural Calamities index	0.500	0.444	0.500	0.457	0.434	0.439	0.457	0.500	0.463	0.472	0.467	
	Stress Index	0.500	0.500	0.500	0.500	0.500	0.500	0.500	0.500	0.500	0.500	0.500	
	Climatic disease index	0.417	0.417	0.417	0.417	0.417	0.417	0.417	0.417	0.417	0.417	0.417	
Natural Resources	Natural resource availability index	1.000	1.000	1.000	1.000	1.000	1.000	1.000	1.000	1.000	1.000	1.000	0.793
	Natural resource collection index	1.000	1.000	1.000	1.000	1.000	1.000	1.000	1.000	1.000	1.000	1.000	
	Natural resource degradation index	1.000	1.000	1.000	1.000	1.000	1.000	1.000	1.000	1.000	1.000	1.000	
	Natural water quality index	0.500	0.500	0.500	0.416	0.500	0.392	0.416	0.500	0.455	0.416	0.459	
	Fodder strategy index	0.400	0.500	0.500	0.500	0.500	0.478	0.500	0.500	0.500	0.500	0.488	
	Fuelwood strategy index	0.308	1.000	1.000	1.000	1.000	0.367	1.000	1.000	1.000	0.455	0.813	
Health	Health worker index	0.267	0.350	0.350	0.267	0.225	0.267	0.350	0.350	0.350	0.267	0.304	0.411
	Health facility accessibility index	0.443	0.386	0.786	0.543	0.286	0.700	0.643	0.500	0.400	0.486	0.517	
Infrastructure	Connectivity index	0.167	0.267	0.667	0.250	0.417	0.250	0.417	0.417	0.333	0.417	0.360	0.427
	Market accessibility index	0.000	0.667	1.000	0.000	0.667	1.000	0.667	1.000	0.333	0.667	0.600	
	Bank accessibility index	0.286	0.571	0.571	0.286	0.571	1.000	0.286	0.000	0.000	0.571	0.414	
	Government resource index	0.333	0.333	0.333	0.333	0.333	0.333	0.333	0.333	0.333	0.333	0.333	

Annex 4: Indexed components and sub-components for vulnerability and capacity assessment of Jaora block

Components	Sub-Components	Rojana	Banbara	Bhuteda	Uplai	Kerwasa	Dodiyana	Udyakhedi	Patakhedi	Rewas	Badodiya	Jaora	Ja_VCA
Socio-Economic Profile	Family dependency index	0.550	0.400	0.392	0.392	0.375	0.342	0.327	0.300	0.528	0.490	0.410	0.209
	Profession diversity index	0.250	0.333	0.250	0.250	0.250	0.200	0.333	0.200	0.200	0.200	0.247	
	Poverty line index	0.200	0.100	0.200	0.300	0.100	0.100	0.000	0.200	0.200	0.100	0.150	
	House type diversity index	0.250	0.256	0.263	0.277	0.277	0.277	0.270	0.321	0.285	0.256	0.273	
	Asset index	0.106	0.125	0.160	0.160	0.166	0.158	0.188	0.178	0.152	0.134	0.153	
	Livestock index	0.185	0.149	0.208	0.217	0.185	0.296	0.238	0.169	0.102	0.161	0.191	
	Migration index	0.000	0.300	0.000	0.000	0.000	0.000	0.000	0.000	0.000	0.000	0.030	
	Electricity index	0.083	0.000	0.000	0.000	0.000	0.000	0.000	0.000	0.000	0.000	0.008	
	Sanitation index	0.200	0.500	0.500	0.200	0.200	0.300	0.300	0.200	0.300	0.000	0.270	
	Stability index	0.300	0.200	0.500	0.600	0.200	0.500	0.300	0.500	0.300	0.200	0.360	
Social Network	Assistance index	0.333	0.358	0.351	0.375	0.298	0.318	0.339	0.354	0.377	0.328	0.343	0.367
	Money transaction index	0.492	0.370	0.461	0.527	0.459	0.366	0.301	0.298	0.331	0.294	0.390	
Education	Youth education index	0.000	0.052	0.062	0.000	0.077	0.000	0.059	0.000	0.411	0.167	0.083	0.151
	Household education index	0.111	0.236	0.117	0.103	0.150	0.175	0.175	0.050	0.452	0.260	0.183	
	Household decision index	0.333	0.133	0.233	0.266	0.166	0.133	0.100	0.066	0.166	0.266	0.186	
Agriculture	Agricultural equipments index	0.148	0.117	0.125	0.148	0.125	0.125	0.129	0.339	0.235	0.222	0.171	0.188
	Agricultural land index	0.157	0.061	0.070	0.416	0.010	0.013	0.000	0.247	0.054	0.037	0.106	
	Cropping intensity index	0.283	0.122	0.210	0.167	0.100	0.100	0.000	0.396	0.216	0.148	0.174	
	Irrigation index	0.138	0.145	0.130	0.165	0.130	0.138	0.152	0.154	0.152	0.154	0.146	
	Irrigation strategy index	0.256	0.370	0.263	0.250	0.263	0.384	0.370	0.380	0.384	0.333	0.325	
	Agricultural strategy index	0.166	0.166	0.228	0.243	0.192	0.181	0.166	0.300	0.172	0.250	0.206	
Food	Food sufficiency index	0.225	0.229	0.232	0.263	0.225	0.235	0.225	0.253	0.225	0.233	0.235	0.147
	Nutritious food index	0.051	0.056	0.057	0.053	0.066	0.061	0.058	0.072	0.062	0.063	0.060	
Water	Portable water Accessibility Index	0.081	0.306	0.111	0.400	0.200	0.083	0.375	0.360	0.135	0.169	0.222	0.134

	Portable water availability index	0.100	0.276	0.071	0.000	0.000	0.000	0.332	0.197	0.085	0.000	0.106	
	Portable water collection index	0.000	0.500	0.000	0.000	0.000	0.000	0.092	0.141	0.000	0.000	0.073	
Weather Perception and Calamities	Weather Forecast index	0.265	0.327	0.417	0.289	0.359	0.283	0.291	0.339	0.291	0.313	0.317	0.429
	Summer temperature perception index	0.500	0.500	0.467	0.477	0.446	0.458	0.488	0.475	0.455	0.477	0.474	
	Rainfall perception index	0.500	0.500	0.337	0.401	0.340	0.389	0.500	0.335	0.360	0.428	0.409	
	Winter temperature perception index	0.500	0.500	0.434	0.400	0.476	0.384	0.500	0.391	0.384	0.416	0.439	
	Natural Calamities index	0.500	0.500	0.433	0.484	0.425	0.461	0.500	0.453	0.387	0.500	0.464	
	Stress Index	0.500	0.476	0.500	0.500	0.500	0.450	0.500	0.500	0.400	0.500	0.483	
	Climatic disease index	0.488	0.442	0.435	0.442	0.442	0.366	0.410	0.416	0.358	0.393	0.419	
Natural Resources	Natural resource availability index	0.333	0.333	0.333	0.333	0.333	0.333	0.333	0.333	0.333	0.333	0.333	0.376
	Natural resource collection index	0.500	0.500	0.488	0.500	0.477	0.500	0.500	0.455	0.435	0.500	0.485	
	Natural resource degradation index	0.208	0.179	0.202	0.208	0.196	0.208	0.179	0.185	0.196	0.208	0.197	
	Natural water quality index	0.500	0.500	0.500	0.500	0.500	0.500	0.500	0.500	0.500	0.500	0.500	
	Fodder strategy index	0.344	0.500	0.434	0.416	0.384	0.434	0.370	0.450	0.500	0.500	0.433	
	Fuelwood strategy index	0.256	0.250	0.294	0.357	0.250	0.312	0.250	0.500	0.250	0.357	0.308	
Health	Health worker index	0.200	0.267	0.225	0.267	0.225	0.225	0.225	0.225	0.267	0.267	0.239	0.377
	Health facility accessibility index	0.000	0.143	0.143	0.429	0.643	0.857	1.000	0.857	0.500	0.571	0.514	
Infrastructure	Connectivity index	0.167	0.167	0.417	0.417	0.167	0.167	0.417	0.417	0.417	0.417	0.317	0.295
	Market accessibility index	0.000	0.000	0.000	0.000	0.000	0.000	0.000	0.000	0.000	0.000	0.000	
	Bank accessibility index	0.000	0.143	0.143	0.429	0.643	0.857	1.000	0.857	0.500	0.571	0.514	
	Government resource index	0.333	0.333	0.333	0.333	0.333	0.333	0.417	0.333	0.333	0.417	0.350	

Annex 5: Indexed components and sub-components for vulnerability and capacity assessment of Bajna Block

Components	Sub-Components	Maanpur	Chotiya Bawdi	Gadika-tara Khurd	Manda-liya	Badan Khurd	Karbal Khora	Lamb-isada	Lalpur	Mahu-ripada	Jharaniya Ukala	Bajna	Bj_VCA
Socio-Economic Profile	Family dependency index	0.472	0.400	0.464	0.530	0.547	0.569	0.456	0.527	0.540	0.617	0.512	0.594
	Profession diversity index	0.500	0.200	0.333	0.250	0.333	0.333	0.250	0.333	0.333	0.333	0.320	
	Poverty line index	0.900	0.900	0.900	1.000	0.900	1.000	1.000	1.000	1.000	1.000	0.960	
	House type diversity index	0.500	0.466	0.500	0.500	0.500	0.500	0.416	0.475	0.325	0.275	0.446	
	Asset index	0.256	0.279	0.500	0.263	0.400	0.400	0.295	0.352	0.333	0.280	0.336	
	Livestock index	0.417	0.358	0.202	0.170	0.241	0.205	0.199	0.209	0.195	0.197	0.239	
	Migration index	1.000	1.000	1.000	0.800	1.000	1.000	0.900	0.900	1.000	0.900	0.950	
	Electricity index	0.312	0.250	0.250	0.000	0.208	0.250	0.000	0.125	0.125	0.208	0.173	
	Sanitation index	1.000	1.000	1.000	1.000	1.000	1.000	1.000	1.000	1.000	1.000	1.000	
	Stability index	1.000	1.000	1.000	1.000	1.000	1.000	1.000	1.000	1.000	1.000	1.000	
Social Network	Assistance index	0.423	0.455	0.463	0.477	0.385	0.375	0.370	0.412	0.371	0.339	0.407	0.483
	Money transaction index	0.772	0.693	0.688	0.426	0.369	0.558	0.190	0.737	0.690	0.472	0.560	
Education	Youth education index	0.400	1.000	0.613	0.452	0.577	0.697	0.516	0.769	0.500	0.438	0.596	0.543
	Household education index	0.789	0.780	0.710	0.692	0.604	0.537	0.718	0.455	0.641	0.671	0.660	
	Household decision index	0.533	0.067	0.433	0.333	0.500	0.267	0.367	0.333	0.433	0.467	0.373	
Agriculture	Agricultural equipments index	1.000	1.000	1.000	0.700	1.000	1.000	0.800	1.000	1.000	0.900	0.940	0.605
	Agricultural land index	1.000	0.908	0.888	0.889	0.730	0.848	0.606	0.847	0.822	0.747	0.829	
	Cropping intensity index	1.000	0.550	0.874	0.750	0.444	0.643	0.381	0.820	0.783	0.661	0.691	
	Irrigation index	1.000	0.453	0.647	0.497	0.373	0.354	0.220	0.525	0.376	0.419	0.486	
	Irrigation strategy index	0.500	0.347	0.500	0.476	0.500	0.500	0.500	0.500	0.500	0.642	0.497	
	Agricultural strategy index	0.167	0.273	0.167	0.167	0.167	0.167	0.167	0.167	0.167	0.258	0.186	
Food	Food sufficiency index	0.310	0.275	0.269	0.278	0.351	0.339	0.241	0.281	0.283	0.297	0.292	0.175
	Nutritious food index	0.054	0.052	0.057	0.060	0.056	0.060	0.056	0.056	0.051	0.068	0.057	
Water	Portable water Accessibility Index	0.226	0.250	0.214	0.218	0.270	0.321	0.208	0.296	0.274	0.338	0.261	0.300

	Portable water availability index	0.151	0.291	0.172	0.291	0.201	0.207	0.038	0.269	0.229	0.217	0.207	
	Portable water collection index	0.373	0.345	0.344	0.413	0.534	0.662	0.153	0.718	0.515	0.258	0.432	
Weather Perception and Calamities	Weather Forecast index	0.389	0.416	0.374	0.385	0.394	0.428	0.467	0.428	0.465	0.361	0.411	0.480
	Summer temperature perception index	0.500	0.500	0.500	0.500	0.500	0.500	0.500	0.500	0.500	0.446	0.495	
	Rainfall perception index	0.444	0.476	0.434	0.476	0.473	0.500	0.500	0.500	0.500	0.500	0.480	
	Winter temperature perception index	0.333	0.500	0.478	0.492	0.491	0.500	0.500	0.500	0.500	0.484	0.478	
	Natural Calamities index	0.500	0.476	0.500	0.500	0.500	0.500	0.500	0.500	0.500	0.500	0.498	
	Stress Index	0.500	0.500	0.500	0.500	0.500	0.500	0.500	0.500	0.500	0.500	0.500	
	Climatic disease index	0.500	0.416	0.870	0.422	0.486	0.500	0.475	0.428	0.477	0.416	0.499	
Natural Resources	Natural resource availability index	0.351	0.377	0.392	0.408	0.394	0.399	0.384	0.358	0.358	0.375	0.380	0.467
	Natural resource collection index	0.416	0.416	0.455	0.416	0.475	0.458	0.450	0.428	0.458	0.351	0.432	
	Natural resource degradation index	0.458	0.437	0.458	0.458	0.458	0.458	0.458	0.458	0.458	0.422	0.452	
	Natural water quality index	0.500	0.500	0.500	0.500	0.500	0.500	0.500	0.500	0.500	0.500	0.500	
	Fodder strategy index	0.500	0.455	0.500	0.500	0.500	0.500	0.500	0.500	0.500	0.400	0.485	
	Fuelwood strategy index	1.000	0.500	0.500	0.500	0.500	0.500	0.500	0.500	0.500	0.500	0.550	
Health	Health worker index	0.267	0.600	0.350	0.267	0.350	0.350	0.267	0.350	0.267	0.600	0.367	0.396
	Health facility accessibility index	0.375	0.375	0.000	0.625	0.375	0.625	1.000	0.000	0.125	0.750	0.425	
Infrastructure	Connectivity index	0.375	0.375	0.375	0.375	0.375	0.375	0.375	0.375	0.375	0.375	0.375	0.359
	Market accessibility index	0.500	0.000	0.100	0.000	0.200	0.200	1.000	0.200	0.300	0.100	0.260	
	Bank accessibility index	0.375	0.375	0.000	0.125	0.375	0.625	1.000	0.000	0.125	0.750	0.375	
	Government resource index	0.417	0.500	0.333	0.417	0.417	0.417	0.417	0.500	0.500	0.333	0.425	

Annex 6: Vulnerability Score of Villages, Blocks and Districts of Malwa Region (n=400)

District	Block	Village	Exposure	Sensitivity	Adaptive Capacity	Vulnerability= (Exposure+ Sensitivity)- Adaptive Capacity	Vulnerability Score of Blocks
Dhar	**Dhar**	Antrai	0.549	0.238	0.263	0.524	0.453
		Saktali	0.517	0.207	0.311	0.413	
		Lohari Bujurg	0.542	0.248	0.338	0.452	
		Baggad	0.516	0.221	0.352	0.386	
		Samar	0.540	0.276	0.291	0.525	
		Masawda	0.504	0.229	0.308	0.426	
		Sukheda	0.550	0.228	0.267	0.510	
		Nekpur	0.512	0.146	0.283	0.374	
		Sirsoda	0.540	0.181	0.284	0.437	
		Tornod	0.543	0.213	0.270	0.486	
	Gandhwani	Baleri	0.594	0.504	0.462	0.636	0.693
		Gungi devi	0.655	0.424	0.484	0.595	
		Jalokya	0.660	0.593	0.581	0.673	
		Karcha	0.635	0.607	0.512	0.731	
		Kelikhurd	0.650	0.563	0.491	0.723	
		Loharpura	0.577	0.651	0.543	0.685	
		Maalgarh	0.635	0.609	0.564	0.680	
		Machalli	0.660	0.582	0.552	0.689	
		Motimali	0.647	0.717	0.509	0.855	
		Panchpipaliya	0.600	0.590	0.531	0.659	
Ratlam	**Jaora**	Rojana	0.427	0.212	0.180	0.459	0.338
		Banbara	0.432	0.224	0.227	0.429	
		Bhuteda	0.405	0.167	0.226	0.346	
		Uplai	0.418	0.171	0.266	0.323	
		Kerwasa	0.397	0.118	0.226	0.289	
		Dodiyana	0.400	0.113	0.251	0.262	

		Udyakhedi	0.419	0.135	0.283	0.271	
		Patakhedi	0.416	0.208	0.287	0.337	
		Rewas	0.380	0.197	0.256	0.321	
		Badodiya	0.426	0.149	0.236	0.339	
	Bajna	Maanpur	0.500	0.754	0.509	0.745	0.643
		ChotiyaBawdi	0.463	0.599	0.486	0.576	
		GadikataraKhurd	0.507	0.676	0.449	0.735	
		Mandaliya	0.473	0.645	0.421	0.696	
		BadanKhurd	0.481	0.578	0.470	0.589	
		KarbalKhora	0.485	0.629	0.502	0.612	
		Lambisada	0.481	0.499	0.484	0.496	
		Lalpur	0.473	0.674	0.458	0.688	
		Mahuripada	0.479	0.655	0.451	0.683	
		Jharaniya Ukala	0.450	0.636	0.479	0.606	

Annex 7: Mean (± SD) of Tribal (Gandhwani and Bajna) and Non Tribal (Dhar and Bajna) Blocks (n=400)

Components	Indices	Non tribal ± SD	Tribal ± SD	Total ± SD
Socio-Economic Profile	Family dependency index	0.404 ± 0.098	0.499 ± 0.094	0.451 ± 0.106
	Profession diversity index	0.242 ± 0.045	0.351 ± 0.085	0.297 ± 0.087
	Poverty line index	0.215 ± 0.149	0.980 ± 0.041	0.598 ± 0.402
	House type diversity index	0.296 ± 0.048	0.456 ± 0.068	0.376 ± 0.100
	Asset index	0.187 ± 0.086	0.479 ± 0.190	0.333 ± 0.207
	Livestock index	0.245 ± 0.091	0.220 ± 0.066	0.232 ± 0.079
	Migration index	0.035 ± 0.093	0.802 ± 0.178	0.418 ± 0.413
	Electricity index	0.221 ± 0.239	0.457 ± 0.310	0.339 ± 0.298
	Sanitation index	0.238 ± 0.150	0.965 ± 0.068	0.602 ± 0.385
	Stability index	0.374 ± 0.110	1.000 ± 0.000	0.687 ± 0.326
Social Network	Assistance index	0.280 ± 0.082	0.530 ± 0.264	0.405 ± 0.230
	Money transaction index	0.367 ± 0.081	0.559 ± 0.145	0.463 ± 0.151
Education	Youth education index	0.076 ± 0.107	0.529 ± 0.178	0.302 ± 0.271
	Household education index	0.196 ± 0.119	0.668 ± 0.103	0.432 ± 0.263
	Household decision index	0.172 ± 0.085	0.384 ± 0.107	0.278 ± 0.144
Agriculture	Agricultural equipments index	0.201 ± 0.113	0.848 ± 0.223	0.525 ±0.371
	Agricultural land index	0.100 ± 0.117	0.802 ± 0.161	0.451 ± 0.382
	Cropping intensity index	0.187 ± 0.140	0.700 ± 0.178	0.443 ± 0.304
	Irrigation index	0.199 ± 0.057	0.505 ± 0.212	0.352 ± 0.218
	Irrigation strategy index	0.332 ± 0.052	0.496 ± 0.052	0.414 ± 0.097
	Agricultural strategy index	0.155 ± 0.063	0.149 ± 0.054	0.152 ± 0.058
Food	Food sufficiency index	0.238 ± 0.016	0.281 ± 0.030	0.259 ±0.033
	Nutritious food index	0.094 ± 0.123	0.073 ± 0.019	0.084 ± 0.088
Water	Portable water Accessibility Index	0.256 ± 0.139	0.315 ± 0.108	0.285 ± 0.126
	Portable water availability index	0.124 ± 0.148	0.169 ± 0.090	0.146 ± 0.123
	Portable water collection index	0.193 ± 0.187	0.580 ± 0.210	0.386 ± 0.277
Weather Perception and Calamities	Weather Forecast index	0.323 ± 0.052	0.394 ± 0.051	0.359 ± 0.062
	Summer temperature perception index	0.476 ± 0.019	0.496 ± 0.012	0.486 ± 0.019
	Rainfall perception index	0.410 ± 0.067	0.475 ± 0.028	0.443 ± 0.060
	Winter temperature perception index	0.441 ± 0.049	0.472 ± 0.045	0.456 ± 0.049
	Natural Calamities index	0.464 ± 0.038	0.482 ± 0.024	0.473 ± 0.033
	Stress Index	0.483 ± 0.033	0.500 ± 0.000	0.491 ± 0.024
	Climatic disease index	0.418 ± 0.039	0.458 ± 0.102	0.438 ± 0.079
Natural resource	Natural resource availability index	0.417 ± 0.086	0.690 ± 0.319	0.553 ± 0.269
	Natural resource collection index	0.743 ± 0.264	0.716 ± 0.292	0.729 ± 0.275
	Natural resource degradation index	0.598 ± 0.412	0.726 ± 0.281	0.662 ± 0.354
	Natural water quality index	0.470 ± 0.053	0.480 ± 0.037	0.475 ± 0.046
	Fodder strategy index	0.434 ± 0.054	0.487 ± 0.032	0.460 ± 0.051
	Fuel-wood strategy index	0.306 ± 0.078	0.682 ± 0.271	0.494 ± 0.274
Health	Health worker index	0.239 ± 0.024	0.336 ± 0.100	0.287 ± 0.087
	Health facility accessibility index	0.492 ± 0.317	0.471 ± 0.255	0.482 ± 0.284
Infrastructure	Connectivity index	0.334 ± 0.115	0.368 ± 0.097	0.351 ± 0.106
	Market accessibility index	0.500 ± 0.513	0.430 ± 0.375	0.465 ± 0.445
	Bank accessibility index	0.333 ± 0.303	0.395 ± 0.311	0.364 ± 0.305

Annex 8: Independent Samples T-Test for Indices of Tribal (Gandhwani and Bajna) and Non Tribal (Dhar and Bajna) blocks and Ratlam and Dhar (n=400)

Components	Indexes	Tribal and Non Tribal P-Value (<0.05)	Dhar and Ratlam P-Value (<0.05)
Socio-Economic Profile	Family dependency index	0.003	0.580
	Profession diversity index	0.000	0.326
	Poverty line index	0.000	0.510
	House type diversity index	0.000	0.302
	Asset index	0.000	0.005
	Livestock index	0.316	0.171
	Migration index	0.000	0.279
	Electricity index	0.010	0.000
	Sanitation index	0.000	0.589
	Stability index	0.000	0.892
Social Network	Assistance index	0.000	0.421
	Money transaction index	0.000	0.624
Education	Youth education index	0.000	0.391
	Household education index	0.000	0.803
Agriculture	Household decision index	0.000	0.942
	Agricultural equipments index	0.000	0.604
	Agricultural land index	0.000	0.791
	Cropping intensity index	0.000	0.822
	Irrigation index	0.000	0.307
	Irrigation strategy index	0.000	0.837
	Agricultural strategy index	0.770	0.000
Food	Food sufficiency index	0.000	0.441
	Nutritious food index	0.456	0.070
Water	Portable water accessibility Index	0.140	0.027
	Portable water availability index	0.253	0.617
	Portable water collection index	0.000	0.001
Weather Perception and Calamities	Weather Forecast index	0.000	0.585
	Summer temperature perception index	0.000	0.700
	Rainfall perception index	0.000	0.829
	Winter temperature perception index	0.046	0.832
	Natural Calamities index	0.083	0.132
	Stress Index	0.022	1.000
	Climatic disease index	0.112	0.093
Natural resource	Natural resource availability index	0.001	0.000
	Natural resource collection index	0.764	0.000
	Natural resource degradation index	0.259	0.000
	Natural water quality index	0.489	0.000
	Fodder strategy index	0.001	0.916
	Fuel-wood strategy index	0.000	0.134

Health	Health worker index	0.000	0.262
	Health facility accessibility index	0.818	0.793
Infrastructure	Connectivity index	0.318	0.788
	Market accessibility index	0.626	0.000
	Bank accessibility index	0.529	0.094
	Government resource index	0.019	0.000

Annex. 9: Variability within and between the Tribal and Non-Tribal community of Malwa Region

			Mean Square	F	Sig.
Family dependency index Tribal nontribal	Between Groups	(Combined)	.091	9.946	.003
	Within Groups		.009		
	Total				
Profession diversity index Tribal nontribal	Between Groups	(Combined)	.119	25.839	.000
	Within Groups		.005		
	Total				
Poverty line index Tribal nontribal	Between Groups	(Combined)	5.848	488.576	.000
	Within Groups		.012		
	Total				
House type diversity index Tribal nontribal	Between Groups	(Combined)	.258	74.384	.000
	Within Groups		.003		
	Total				
Asset index Tribal nontribal	Between Groups	(Combined)	.852	39.204	.000
	Within Groups		.022		
	Total				
Livestock index Tribal nontribal	Between Groups	(Combined)	.007	1.037	.315
	Within Groups		.006		
	Total				
Migration index Tribal nontribal	Between Groups	(Combined)	5.882	290.053	.000
	Within Groups		.020		
	Total				
Electricity index Tribal nontribal	Between Groups	(Combined)	.559	7.307	.010
	Within Groups		.077		
	Total				
Sanitation index Tribal nontribal	Between Groups	(Combined)	5.277	390.896	.000
	Within Groups		.014		
	Total				
Stability index Tribal nontribal	Between Groups	(Combined)	3.915	651.540	.000
	Within Groups		.006		
	Total				
Assistance index Tribal nontribal	Between Groups	(Combined)	.623	16.319	.000
	Within Groups		.038		
	Total				
Money transaction index Tribal nontribal	Between Groups	(Combined)	.369	26.800	.000
	Within Groups		.014		
Money transaction index Tribal nontribal	Total				
Youth education index Tribal nontribal	Between Groups	(Combined)	2.053	95.424	.000
	Within Groups		.022		
	Total				
Household education index Tribal nontribal	Between Groups	(Combined)	2.231	179.773	.000
	Within Groups		.012		
	Total				
Household decision index Tribal nontribal	Between Groups	(Combined)	.451	48.300	.000
	Within Groups		.009		
	Total				

Agricultural equipments index Tribal nontribal	Between Groups	(Combined)	4.178	133.869	.000
	Within Groups		.031		
	Total				
Agricultural land index Tribal nontribal	Between Groups	(Combined)	4.932	250.166	.000
	Within Groups		.020		
	Total				
Cropping intensity index Tribal nontribal	Between Groups	(Combined)	2.632	102.930	.000
	Within Groups		.026		
	Total				
Irrigation index Tribal nontribal	Between Groups	(Combined)	.936	38.941	.000
	Within Groups		.024		
	Total				
Irrigation strategy index Tribal nontribal	Between Groups	(Combined)	.267	97.773	.000
	Within Groups		.003		
	Total				
Agricultural strategy index Tribal nontribal	Between Groups	(Combined)	.000	.086	.770
	Within Groups		.003		
	Total				
Food sufficiency index Tribal nontribal	Between Groups	(Combined)	.019	32.916	.000
	Within Groups		.001		
	Total				
Nutritious food index Tribal nontribal	Between Groups	(Combined)	.004	.567	.456
	Within Groups		.008		
	Total				
Portable water Accessibility Index Tribal nontribal	Between Groups	(Combined)	.035	2.266	.140
Portable water Accessibility Index Tribal nontribal	Within Groups		.015		
	Total				
Portable water availability index Tribal nontribal	Between Groups	(Combined)	.020	1.350	.253
	Within Groups		.015		
	Total				
Portable water collection index Tribal nontribal	Between Groups	(Combined)	1.503	38.159	.000
	Within Groups		.039		
	Total				
Weather Forecast index Tribal nontribal	Between Groups	(Combined)	.051	19.282	.000
	Within Groups		.003		
	Total				
Summer temperature perception index Tribal nontribal	Between Groups	(Combined)	.004	15.935	.000
	Within Groups		.000		
	Total				
Rainfall perception index Tribal nontribal	Between Groups	(Combined)	.043	16.231	.000
	Within Groups		.003		
	Total				
Winter temperature perception index Tribal nontribal	Between Groups	(Combined)	.010	4.276	.046
	Within Groups		.002		
	Total				
Natural Calamities index Tribal nontribal	Between Groups	(Combined)	.003	3.172	.083
	Within Groups		.001		
	Total				

Stress Index Tribal nontribal	Between Groups	(Combined)	.003	5.725	.022
	Within Groups		.001		
	Total				
Climatic disease index Tribal nontribal	Between Groups	(Combined)	.016	2.652	.112
	Within Groups		.006		
	Total				
Natural resource availability index Tribal nontribal	Between Groups	(Combined)	.747	13.730	.001
	Within Groups		.054		
	Total				
Natural resource collection index Tribal nontribal	Between Groups	(Combined)	.007	.091	.764
	Within Groups		.078		
	Total				
Natural resource degradation index Tribal nontribal	Between Groups	(Combined)	.163	1.311	.259
	Within Groups		.124		
	Total				
Natural water quality index Tribal nontribal	Between Groups	(Combined)	.001	.487	.489
	Within Groups		.002		
	Total				
Fodder strategy index Tribal nontribal	Between Groups	(Combined)	.028	14.292	.001
	Within Groups		.002		
	Total				
Fuelwood strategy index Tribal nontribal	Between Groups	(Combined)	1.407	35.362	.000
	Within Groups		.040		
	Total				
Health worker index Tribal nontribal	Between Groups	(Combined)	.093	17.356	.000
	Within Groups		.005		
	Total				
Health facility accessibility index Tribal nontribal	Between Groups	(Combined)	.004	.053	.818
	Within Groups		.083		
	Total				
Connectivity index Tribal nontribal	Between Groups	(Combined)	.012	1.024	.318
	Within Groups		.011		
	Total				
Market accessibility index Tribal nontribal	Between Groups	(Combined)	.049	.242	.625
	Within Groups		.202		
	Total				
Bank accessibility index Tribal nontribal	Between Groups	(Combined)	.038	.403	.529
	Within Groups		.094		
	Total				
Government resource index Tribal nontribal	Between Groups	(Combined)	.014	6.036	.019
	Within Groups		.002		
	Total				

Annex. 10: Percentage variance (Communalities) explained by factors.

S.No.	Factors	Extraction
1	Socio Economic	.930
2	Agriculture	.866
3	Education	.849
4	Health	.842
5	Infrastructure	.833
6	Weather Perception and Calamities	.807
7	Natural Resource	.788
8	Social Network	.729
9	Water	.602
10	Food	.374

Annex. 11: Total Variance Explained for Dhar and Ratlam District of Malwa Region

Component	Initial Eigen values			Extraction Sums of Squared Loadings			Rotation Sums of Squared Loadings		
	Total	% of Variance	Cumulative %	Total	% of Variance	Cumulative %	Total	% of Variance	Cumulative %
1	4.917	49.165	49.165	4.917	49.165	49.165	4.587	45.872	45.872
2	1.610	16.101	65.266	1.610	16.101	65.266	1.582	15.815	61.687
3	1.094	10.941	76.207	1.094	10.941	76.207	1.452	14.521	76.207
4	.981	9.808	86.016						
5	.494	4.939	90.955						
6	.399	3.987	94.942						
7	.219	2.187	97.128						
8	.167	1.672	98.800						
9	.091	.914	99.714						
10	.029	.286	100.000						

Annex. 12: Component Score Coefficient Matrix for Dhar and Ratlam District of Malwa Region

S. No.		Component		
		1	2	3
1.	Socio Economic Profile	.205	-.008	.024
2.	Social Network	.216	-.126	-.112
3.	Education	.236	-.177	.044
4.	Agriculture	.239	-.180	.041
5.	Food	-.103	.478	-.137
6.	Water	.104	.178	.021
7.	Weather Perception and Calamities	.129	.211	-.066
8.	Natural Resources	.006	.455	.007
9.	Health	.049	-.308	.674
10.	Infrastructure	-.099	.277	.478

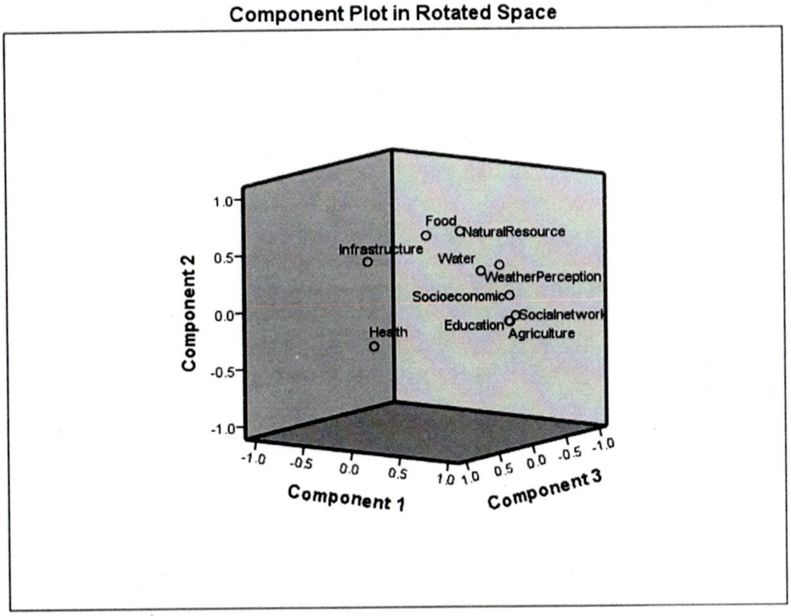

Annex. 13: Total Variance Explained for Tribal blocks of of Dhar and Ratlam District Malwa Region

Component	Initial Eigen values			Extraction Sums of Squared Loadings			Rotation Sums of Squared Loadings		
	Total	% of Variance	Cumulative %	Total	% of Variance	Cumulative %	Total	% of Variance	Cumulative %
1	3.339	33.386	33.386	3.339	33.386	33.386	3.245	32.449	32.449
2	2.055	20.554	53.940	2.055	20.554	53.940	1.763	17.626	50.075
3	1.467	14.666	68.606	1.467	14.666	68.606	1.759	17.591	67.666
4	1.050	10.500	79.107	1.050	10.500	79.107	1.144	11.441	79.107
5	.639	6.385	85.492						
6	.533	5.333	90.825						
7	.458	4.579	95.404						
8	.234	2.336	97.740						
9	.123	1.226	98.966						
10	.103	1.034	100.000						

Annex. 14: Component Score Coefficient Matrix for Tribal blocks of Dhar and Ratlam District of Malwa Region

	Component			
	1	2	3	4
Socioeconomic	.276	-.099	.021	-.120
Social network	.125	-.014	.432	.158
Education	-.063	.182	.522	-.017
Agriculture	-.011	-.223	.305	-.193
Food	-.077	-.009	-.003	.864
Water	.203	-.052	-.066	.119
Weather Perception	.298	.045	-.070	-.128
Natural Resource	.287	.107	.056	-.034
Health	-.078	.493	.142	.149
Infrastructure	.076	.482	.005	-.220

Component Plot in Rotated Space

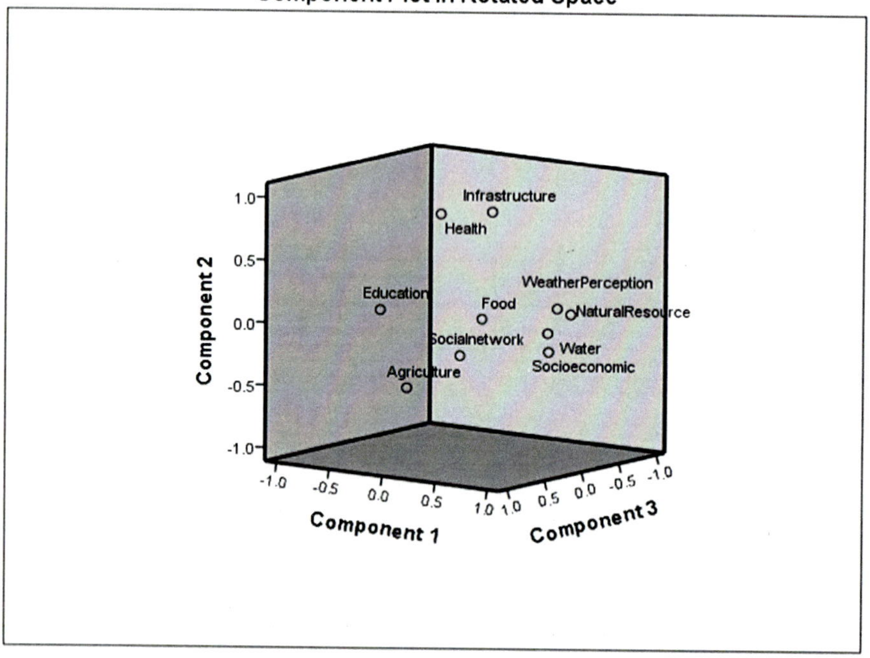